SCOPE

In this series monographs of general biological interest are published.

Ecology, fully recognised as a branch of science since the beginning of this century, gains steadily in importance in all branches of its application. It is a decisive factor for the future of humanity, for the production of food and especially for the maintenance of that fertility of the soil which we have received from our ancestors and which we have the duty to transmit undisturbed to our descendents. This ecological conception must replace the pure technocratic utilization of Nature's wealthes, taking into consideration only the benefit of to-day and leaving only man-made deserts to future generations.

Ecology on this new background has been a basis for research in a few countries like Finland, Australia a.o. who visualize its decisive importance. Therefore the volumes of this series dedicated to ecological research in these countries are of much more than local interest: they offer a new and promising conception of the world.

Other fields to be dealt with in the *Monographiae Biologicae* are problems of comparative and general physiology, of genetics of all organisms.

Due to technical reasons the numeration of the volumes started with Volume V as a continuation of the earlier journal *Physiologia comparata et Oecologia*, **of which four volumes have been published.**

In future, as a rule only volumes dedicated to special topics will be published.

EDITOR

WALTER W. WEISBACH
13, van Stolkweg
The Hague, The Netherlands

ISBN 978-94-017-6908-2   ISBN 978-94-017-7030-9 (eBook)
DOI 10.1007/978-94-017-7030-9

MONOGRAPHIAE BIOLOGICAE

EDITORES

F. S. BODENHEIMER

Jerusalem

W. W. WEISBACH

Den Haag

VOLUMEN V

SPRINGER-SCIENCE+BUSINESS MEDIA, B.V.

VOLUME V

Fasc. 1: pp.   1– 56; edit. 18–12–1957
Fasc. 2: pp.   57–162; edit. 18–12–1958
Fasc. 3: pp. 163–284; edit. 10– 4–1962

ISBN 978-94-017-6908-2   ISBN 978-94-017-7030-9 (eBook)
DOI 10.1007/978-94-017-7030-9

Copyright 1962 by Springer Science+Business Media Dordrecht

Zuid-Nederlandsche Drukkerij N.V. – 's-Hertogenbosch

CONTENTS

Monographiae Biologicae start with Vol. V as continuation of the journal Physiologia comparata et Oecologia, of which Vols. I to IV were published 1948 to 1957.

BIOCHEMICAL ASPECTS
OF
HUMAN MALNUTRITION
IN THE TROPICS

BY

OLUMBE BASSIR

Head of the Biochemistry Laboratory, University
College, Ibadan; Editor of the "West African
Journal of Biological Chemistry"; Formerly,
Nuffield Fellow at the Human Nutrition
Research Unit of the Medical Research
Council, London; At one time, Senior
Clinical Biochemist, Westwood Hospital,
Beverley, England.

SPRINGER-SCIENCE+BUSINESS MEDIA, B.V. - 1962

To the memory of my mother

CONTENTS

CHAPTER I

GEOGRAPHY OF MALNUTRITION

The Environment

For the purpose of this review, the area referred to as the Tropics is that which is encompassed between the latitudes 30° N and 30° S. This is as much an empirical definition as any other. But it has the advantage of dividing the globe into approximately equal halves.

Although the climatic conditions of the tropical areas in each of the continents are fairly uniform, there are distinct pockets in which the climatic profile is more akin to that of temperate zones than tropical. The divergence may be due to the range of mountains or to the presence of large disperses of water, or to the direction of prevailing winds.

The prevailing weather in Kenya at the foot of Mount Kilimanjaro is for most of the year temperate; and the vegetation is more temperate than tropical.

The seasonal characteristics of the tropics are, at any given time, easily discernable from those of the other geographical regions of the world. One of the main features is very heavy rainfall, coupled with a dense tropical vegetation. The periods of maximum and minimum rainfall are usually determined by the migration of the equatorial troughs.

Moving away from the equator, the season changes from a uniform day-to-day rainy state to one in which there are two distinct rainy parts of the year. In the ideal situation, each rainy season lasts for approximately three months with a peak period occurring six weeks after the onset of the rains. Between the rainy seasons comes the period of almost absolute dryness when practically no rain falls for months.

Between the latitude of 15° and 30°, the model is one in which there are two prevailing seasons, one dry and one wet, of approximately equal length. Again, the peak of rainfall varies considerably, depending upon the parallel of latitude and the disposition of the natural features of the landscape.

The Western part of Africa, as far north as Northern Nigeria, illustrates the pattern rather well.

Along the equator, in the Congo Basin, it does rain practically every day, in the afternoon. Around Libreville, there is a short dry season between May and September, and two equally intense rainy periods with maxima in March and November. In Katsina in Northern Nigeria, the rainy season is a brief one lasting from June to October. The heaviest rainfall comes in August. In between the seasonal patterns of Katsina and Libreville lies that of Monrovia, in Liberia, where there is a season of very heavy rainfall, achieving a maximum in July and August.

The yearly average rainfall decreases markedly as one passes from the equatorial zone to the sub-tropics. This is because in most of the continents, high coastal mountains pull the rains down near the sea. By the time the centres of the land masses are reached, the winds have become dry. Hence the desert sub-tropical regions. The exceptions to the general rule are, of course, to be found where there are no high mountain fringes to the land. At sea level, the relative humidity of the tropical atmosphere can be very high indeed. Over 80% at 25° C is the average figure in the western tropical sphere. However, the higher it is above sea level, the less the relative humidity and the ambient temperature. Thus, it is possible to obtain an atmospheric temperature of —70° C and a relative humidity of 48% at a point 6,000 yards above sea level in the Carribean area.

The relative humidity is usually highest at night and lowest at mid-day. This is not the same thing as with vapour pressure which is kept high in the tropics due to constant evaporation from dense forests, wide waterways and so on. It follows that in the rainy season, the relative humidity in the daytime may be as low as 50% although the vapour pressure is high. In the dry hot season, similarly, the relative humidity may only amount to 20%.

From the point of view of the human body, relative humidity is far more important than absolute humidity. Hence the wet-bulb temperature of the environment is nearer physiological than the dry-bulb temperature. When the temperature of the air is approximately the same as that of the body, and the relative humidity is high, evaporation from the skin may be brought virtually to a stand-still. The result is that the physiological mechanisms of the body become incapable of keeping its temperature at 37° C (98.4° F). Hence the feeling of great discomfort which is typical of tropical climates.

In temperate zones, the day temperature in summer may rise to 120° F (49° C) (dry-bulb), but the wet-bulb figure is seldom higher than 80° F (27° C). In Victoria in Australia this type of climate is frequently experienced. But its debilitating effects on humans is far less than a typical 106° F (41° C) (dry-bulb), 98° F (36.6° C) (wet-bulb) of the rain forest climate of Eastern Nigeria.

The temperature of atmospheric air varies with the migration of the tropical trough, being highest in July in the northern hemisphere and in January in the southern hemisphere. The temperature gradient curves for July and January cross each other at a temperature equator which is situated near 5° N. This is due to the fact that the largest land masses in the tropics are situated to the north of the equator. It is also related to the very high temperatures of the northern deserts.

In general, therefore, temperatures in the southern half of the tropics are slightly lower than those felt at corresponding latitudes in the northern half. The difference ranges from 1.4° C to 2.3° C. The seasonal temperature range at 30° N is 13° C, whereas at 30° S it only amounts to 7° C. The mean July average sea level temperature peak is about 30° C, and the corresponding January maximum 28° C.

With regard to the variation of temperature with change in longitude, the most characteristic feature is low mean temperature of the seas west of the American and African continents. These temperature depressions are not improved in winter because of the shapes of the land masses. The force of the cold winds and their effects on oceanic streams are not alleviated in the winter except in the northern hemisphere. The temperature maxima are found over land and there is a cooling off sea-wards except in the Monsoon season over the Indian Ocean. There is a striking temperature gradient at different heights over the tropical oceans. Above and below a point, which is usually situated at a height of about 3 kilometres, temperature lapses occur. The lapse rate is steeper above the point of inversion and reaches 0.8° C per 100 metres in the eastern hemisphere. In the western parts, especially over the equatorial trough, the general picture is of a moist, warm lower layer and a cold, dry upper one.

Although the mean annual temperature ranges are very small, the diurnal variations are usually considerable. In Southern Nigeria there is usually a fall of around 20° F (11° C) in the night. Greater nightly depressions are experienced in other places.

In general, the vegetation of the tropics is "bush". The flora consists of tall large trees reaching far into the sky. Halfway up these shrubs and vines climb to form a canopy. Underneath this are sheltered all kinds of creepers, bulbs, vegetables, mushrooms, ferns, and a myriad of other plants. This is the so-called tropical jungle, impenetrable and mysterious.

Outside the edges of the equatorial trough, the natural vegetation of the tropics is less dense and more distinctive. Again we find a large number of different kinds of trees and shrubs, but these are set out in individual posts by high grass and herbs. Condensation is less suffuse, and the leaves of the plants are less broad.

This is the area of the tropics in which seasonal variations must obviously affect the vegetation. In the dry season, many of the herbs die. The larger plants shed some of their leaves and flower only towards the beginning of the rains.

The grassland which succeeds this park zone to the north is to be found only in those continents where most of the rains falls on the coast.

The effect of these natural vegetations upon the climate as a whole is considerably beyond the limits of the micro-climate. This is not always realized. A few years ago, in an attempt to grow ground-nuts on a large scale in East Africa, large areas of forests were denuded. Consequently, the climate changed, becoming less suitable for both plants and animals.

It is well known that there is a marked difference between the reflectivity of the surfaces of green plants and that of concrete or the bare ground. The difference is based on the wave-length of light considered. In the spectrum above 750 mμ a leafy meadow is more reflective than a dry earth surface. However, within the wave-length range 750—500 mμ the position is strongly reversed.

Radiation transmissivity and light permeability are other factors which, while affecting the climatic environment, are themselves considerably influenced by the natural vegetation.

It is significant that large areas of the tropics are swampy. There, the soil is water-logged but not to the extent of entirely preventing plant growth. In fact, tropical swamps make an excellent habitat, not only for the well-known aquatic plants, but also for many ferns, palms and grasses.

In some places grain plants, like rice, are cultivated almost ex-clusively in swamps. The acidity of the soil water is not too high, and the silt deposits form an ideal base.

In the virgin state, tropical swamps are no less impenetrable than the ordinary tropical forest. Here, also, vines, climbers, etc., combine to bind firmly the larger plants with the small.

The leaves of forest and swamp plants alike control the environ-mental temperature of the tropics as few other factors do. During the day, the temperature of leaf surfaces may be as high as 10° F (5.6° C) above air temperature. At night, the position is reversed. Hence, the soil is protected.

It is easy to see that the combination of high ambient temperature, high rainfall, high vapour pressure and dense vegetation makes for the growth and development of animals. Very little protection is required for the young against the vagaries of the climate. Meta-morphosis is undergone with the minimum of delay. Gestation is no serious problem where food is easily obtained.

The major balancing factor appears to be the fact that larger animals are predatory over smaller and weaker ones. Of course, many millions of animals are destroyed in this way in every continent every year. Rain water washes many others away.

Clearly, any upsetting of this delicate balance, by which the characteristic feature of the fauna is maintained, may have un-predictable consequence. Thus, if the cyclops that feed on mos-quito larvae are destroyed, many more mosquitoes will reach adulthood and many more people will suffer from Malaria, "mutatis mutandis".

Since life in the tropics is shared by man and beast, it is not surprising that many of the animals which comprise the natural fauna are enemies of the human being. In some cases, the life cycle of the animal cannot be completed without the mediation of a human host. This is the situation with many worms. In others, the animal is literally a pest in that the adult requires human tissues for its food. Most frequently, however, the animal pest destroys man's food or his house or his clothes in the process of satisfying the urge to feed or to reproduce.

Even the so-called man-eating lions do not regard the human members of their habitat as staple food. This means that with due deliberation and determination man will, in the end, destroy his animal pests.

Infant Mortality and Diet

Without any qualification, it is possible to say that the highest incidence of infant mortality occurs in the tropics. In Asia and Africa, the rate is five to ten times that of North Western Europe.

Table I gives comparative data for a number of representative

Table I

Comparative Infant Mortality (under one year).

Incidence per 1000 live births

	1954	1955	1956	1957	1958
Union of South Africa White Population	33.3	29.8	31.4	29.1	29.4
Coloured Population	128.7	134.5	137.9	126.7	131.9
Chile	125.1	126.1	116.2	117.2	126.8
Trinidad and Tobago	60.5	67.9	63.9	56.5	62.7
Federation of Malaya	83.1	78.4	75.2	75.5	79.6
England and Wales	25.4	24.9	23.7	23.1	22.5
United States	26.6	26.4	26	26.4	26.9
Australia	22.5	22	21.7	21.4	20.5
Sweden	18.7	17.4	17.3	17.8	15.8

This table was constructed from data supplied in the Statistical Year Book of the United Nations, New York 1959.

countries. WATT[1] has discussed the high mortality of Lagos babies. ONABAMIRO[2] has analyzed the incidence of infant mortality for the British West African territories and showed that the rate for Nigeria and the Gold Coast is approximately 120/1000 of which 3—5% are due to still births. This general picture has remained practically unchanged for the last two decades. The figures for Gambia show a decline from a high incidence of 31% in 1935 to an average of approximately 15% for the last ten years. The case of Sierra Leone is different in that there has been considerable fluctuation in the general trend. However, it appears certain that the average over the last quinquennium is well over 150/1000 births.

Discussing the causes of such extraordinarily high rates of infant mortality, ONABAMIRO[2] puts the practice of indigenous midwifery high on the list. In the absence of exact scientific knowledge of the anatomical presentations of the foetus, indigenous midwives are unlikely to correct the posture of the emerging baby and so to ensure live birth and prevent damage to the baby. The same considerations hold for cases of misplaced placenta and extra-uterine pregnancy. The infection of a baby in the process of being born may also present considerable dangers to the life of the child, and lead to its early death.

It is well known that successful pregnancy and parturition is dependent to a considerable extent on the nutrition of the mother in pregnancy. The diet of pregnant women in tropical countries often lacks the additional protein, minerals and vitamins which ensure that the foetus develops healthily and is not under-sized or frail at birth. A child born of a poorly fed mother may start life at such a disadvantage as to succumb to the first major threat to its life.

Diseases of the pregnant mother may directly affect the life of the foetus and give rise to its death at birth. Among these may be mentioned venereal diseases and tuberculosis. Anemia due to undernutrition is of special importance in this regard. There are cases in which the haemoglobin content of the mother may fall to 20% of normal before parturition. In India, Egypt and West Africa, nutritional anaemia is responsible for a large number of deaths of infants and of their mothers. LAWSON & LISTER[3] conducted observations on a large number of such cases at Ibadan in Nigeria. They record a death rate of these women of over 20% among deaths of pregnant women in hospital, despite blood transfusion and special feeding, in the week prior to the birth of the child. It is not known, in Africa, exactly what proportion of such babies die in infancy, but indirect evidence suggests that it might be in the region of 80%.

Parasitic infection of the tropical child takes a heavy toll of lives. Guinea worm enters the body through drinking water. Malaria organisms are injected by mosquito bites. Germs of various sorts are conveyed to the baby through unhygienic handling, housing and airing. Dysentry, both bacillary and amoebic, is endemic in many parts of the tropics.

It is under the onus of this disease that most children of tropical countries fail to thrive on inadequate breast milk, and poor supplementary foods. Kwashiorkor develops in a large number as a sequel of gross malnutrition and dietary shortage of animal proteins. After a certain stage, the prognosis of Kwashiorkor babies is not at all favourable.

Superstition is one of the curses of most of the economically-underdeveloped countries of Asia and Africa. Many babies die because sickness in the child is attributed to witch-craft. Indigenous medicine men are consulted and prescribe an assortment of curative agents such as amulets, skin ointments, and physical exposure.

Herbal infusions are given to new-born babies in many parts of West Africa. Undoubtedly, these have their value; yet it is quite possible that in many cases severe diarrhoea leading to the death of the infant is caused by over-dosage, if not wrong preparation of the herbs.

The Staple Diet of Young and Old

It is significant that throughout the tropics the pattern of feeding is practically the same. Most of the foodstuffs are cooked before they are eaten. Condiments, e.g., peppers and curry, are lavishly used.

The main staple foods are roots and grains. The roots are often cultivated, but not always. Of the cultivated kinds, yams, cocoyam, sweet potatoes and cassava are of commonest use. The wild roots include certain varieties of yams, certain bulbs and tubers.

Rice and maize are the cereals which are most popular with tropical peoples. Guinea-corn and millet are also extensively used as food. Both in Asia and Africa grains are soaked in water until they ferment or germinate, prior to cooking. The husks are, frequently, not removed until after fermentation is completed. And the starchy part of the food is eaten in the form of a paste or porridge which is prepared with boiling water.

Apart from the peppers, tropical vegetables are cooked in their green state. Sometimes, they are first macerated and washed free of poisonous compounds before being cooked. Almost always, the cooking is thorough. This suggests that most of the ascorbic acid is probably destroyed in the process of cooking. The pH of the vegetable sauces varies from place to place, and from time to time in the same place. Therefore, the degree of destruction of ascorbic acid will also vary.

For the same reason it is probable that the chief sources of aneurin and other B-vitamins in the diet of most tropical peoples are palm wine or the so-called native beers which are brewed from fermented grains.

Fish and meat do not preponderate in African and Asian diets, except in special localities such as the highland of the Kikuyu in East Africa and the inhabitants of the Rivers Division in Southern Nigeria.

Consequently, most indigenous peoples of the tropical countries live on a high carbohydrate-low protein diet. It is possible that the quantities of nuts and leaves consumed by some tribes in India are sufficiently large to provide as much as 50 grams protein a day. But vegetable proteins are known to be deficient in some of the essential amino-acids, of which lysine, methionine and arginine are, perhaps, the most important in tropical nutrition.

The role of the sulphur-bearing amino-acids in the metabolism of tropical nutrition is no doubt accentuated by the need for detoxication of food-poisons which are often eaten. BROADBENT & REIFF[4] have studied the physiology of a dioscorin-like colloid which occurs in some Nigerian edible tubers. The hydrocyanic acid content of some varieties of cassava is excessive and taxes the detoxicating potentialities of the liver of those who feed thereon.

The high carbohydrate content of the staple foods of tropical peoples should make easy the provision of the calorie requirement, but this end is not always achieved. In India and other South-east Asian countries, the density of population is exceedingly high and the food produced is inadequate. Therefore, real hunger, bordering on starvation, is not unknown in those places.

Coupled with an archaic method of agriculture, in which deep-ploughing and general mechanization are still pious hopes, under-nutrition leads directly to under-productivity. Long hours are spent at work, but little fruit is gained in reward.

Among the agricultural peoples of tropical lands some feeding habits are universal. Large and bulky meals are consumed at infrequent intervals. Breakfast is seldom eaten. A mid-day snack and evening feast are all to which the hard-working peasant may aspire.

Taboos and customs are limiting factors which restrict the compass of staple foods in the tropics. Millions of people in India and Pakistan regard the cow as sacred and will not eat beef. Many Burmese are vegetarian and insist upon leaves and nuts as their main diet.

In this unfortunate but logical conflict between common-sense and common practice, many Asians and Africans starve when they need not. Insects[5], snakes, birds and slugs make delicious diets for some indigenous peoples of the Tropics, but their more sophisticated brethren turn their backs to these lowly creatures and feed wholly and solely on tubers and grains.

Harmful as this may be to adults, the effect on infants is cata-strophic. It would appear[6] that African mothers who are town dwellers produce insufficient breast-milk to fully nourish their babies in the later half of the first year of lactation.

Whether it be sophistication or poverty, or both, that is respon-sible for their use of maize or cassava gruels as the sole items of their infants' diet, at the earliest stages, the effects of malnutrition in evoking deficiency diseases and arresting growth are inevitable.

And it is in this same fashion that the fact that most tropical people feed their children on exactly the same dishes, as they them-selves consume, should be viewed.

Productivity and Diet

When the productivity of the various countries of the world, in terms of man-hour, is plotted for agriculture and food (Table II), a definite picture emerges. The countries of Europe and North America, which are the most technically advanced, turn out to be, also, among the most productive both from the point of view of food and industrial products. In terms of national income, per man, the United States are twice as productive as the United Kingdom whose income is approximately £ 250 per man, per year. India, Indonesia and Liberia are about a tenth as productive as Britain, while Nigeria, Kenya and Jamaica are poorer than Britain by a factor of five.

Table II

Index of Food and Agricultural Production (per caput) World
Average 1952/'53 – 1956/'57 = 100

Region	Food		Agriculture	
	1958/'59	1959/'60 (Prelim.)	1958/'59	1959/'60 (Prelim.)
Western Europe	109	113	105	100
North America	110	111	99	99
Africa	105	103	98	95
Far East (Excluding China)	109	113	103	105
Latin America	114	112	105	104
World Average	114	115	106	106

This table has been compiled from statistics supplied by the U.N. Food and Agri-culture Organization in "The State of Food and Agriculture", 1960.

(175)

14

There are hardly any data relating the amount of actual work with food production, since the pattern of work varies from country to country. In terms of hours spent at work, however, the disparity between the productivity of the tropical peoples and that of the Europeans is probably even more startling. For far from enjoying an eight hour day, five and a half days a week, in many tropical countries, the peasants work every day from dawn till dusk, with a short break in the middle of the day.

There are still many places in the tropics where agricultural tools are limited to narrow-blade hoes and matchets. This is certainly true of most of Africa, south of the Sahara. In places like Northern Nigeria and Pakistan where farm animals are used in ploughing, the problem of providing adequate diet for the beast seriously restricts their usefulness on the farm.

Another pattern of agriculture which is characteristic of the tropics is that of 'shifting cultivation'. In this system, the same piece of land is not used for two consecutive crops. Instead, the farmer moves to a new land every year and has the tremendous task of clearing it of all its trees, shrubs and wild animals before he can sow his crops. Obviously, one of the quickest means of achieving the desired end is to cut down the trees, allow them to dry in the sun, and then fire the whole lot. Clearing of the stumps is then comparatively easy. Now, while burning the bush leaves a useful layer of potash on the ground, a great deal of the nitrogenous and carbonaceous properties of a virgin land is lost this way. Useful worms and slugs are destroyed together with harmful snakes and other small animals. Furthermore, it means that four or five times as much land, as people to farm them, must be available if food production is to be maintained at a reasonably high level. This is one reason why there are hunger and deficiency diseases amongst the African peoples of East and Southern Africa where European settlement has robbed the indigenous people of their extra land.

The staple diet of farm labourers and peasants in Africa is mainly vegetable and fruit. Game and other small animals are incorporated into the evening meal, from time to time. But eggs and fresh milk are not in general use. In Hausa country in Northern Nigeria, where the Fulani herdsmen roam, sour milk is an integral part of the diet, although it is doubtful whether the amounts consumed are adequate. Because the agricultural worker lives near to nature, nutritional deficiency diseases are not very common in Africa, unless in those places where special customs (e.g. drinking large amounts of sweet, unfermented palm-wine) lay an additional stress on the body metabolism.

In the Far East, over-population, archaic agricultural methods and religious customs make the peasant's diet not only inadequate but poor in quality. Animal protein is usually lacking. Curries and

(176)

peppers are consumed in considerable quantities; and very little calcium and phosphorus are obtained from dietary milk.

Obesity in the townsfolk is now becoming a problem in many tropical countries. It is very likely that this is due to the fact that people feed on the usual high carbohydrate/low protein diets, but do not perform the hard physical work in which use can be made of the extra calories. In Western Nigeria, where the meals often include a good deal of red palm oil, many unwanted calories are provided in the food. But the manual workers in the towns seldom have their normal B vitamin requirement and they develop deficiency diseases.

These people are not factory workers but porters, road builders, and so on. A common feature of most of the tropical lands is the absence of factories for the production of even consumer goods. There is therefore no easy avenue for productive work for the indigenous population. Machines have to be imported and are therefore expensive. Productivity of all the where-withals of civilised life is low.

Endemic Diseases and Life Span

It is a well-known fact that the longevity of tropical peoples is much lower than that of Europeans and North Americans. The life-span of the average indigenous person living in South-east Asia is still below 40 years, while in the Middle East the length of life of the average person is around 35 years. In West Africa the average life-span is probably over 40 years but well below three score and ten years.

It is necessary to state, however, that in the compounding of

Table III

Reported Cases of Malaria and Fatalities.

Country	Incidence					
	1955		1956		1957	
	Cases	Deaths	Cases	Deaths	Cases	Deaths
French Camerouns	140660	144	269255	85	241575	76
Congo Republic	826392	1855	830347	2230	870283	2177
Guatamala	24010	6831	19818	7348	10220	6570
Iraq	320726	760	217834	434	107204	215

This table is composed from data provided in the W.H.O. Epidemiological and Vital Statistics Report, Geneva, *1959*.

16

these averages little regard is paid to the very high infant mortality which is illustrated in Table I. When allowance is made for this, the averages are raised by about ten years. Even so, they bear no comparison to English and North American means.

The main cause of the high death-rate among tropical peoples is probably endemic diseases, of which Malaria is the most fatal. The incidence in a group of selected tropical countries and the fatality rates are given in Table III. Bearing in mind that the number of cases and deaths reported represent only a fraction of the actual figures, and that these do not include latent and chronic incidences, it can be seen at once how important this particular endemic disease is in reducing the life-span of the Asian and African peoples.

The death-rates from all causes also show the same trend (Table IV). The highest occur in the tropical countries of Asia and South America and the lowest in Scandinavia and North America.

Apart from Malaria, yaws, plague and sleeping sickness help to increase the annual death-roll of the tropical peoples. Leprosy and Pellegra are debilitating and indirectly contribute to the high mortality rate. As late as 1954, 2714 cases of pellagra were reported from Ethiopia, of whom 53 died. In the same year, 1008 cases were notified

Table IV

Deaths due to all causes

Country	Mortality rates			
	1958		1959	
	Popula-tion (1000)	% deaths	Popula-tion	% deaths
Tunisia	3852	1.32	3880	1.18
India	397540	1.19	402750	1.21
Chili	7298	1.21	7465	1.25
Venezuela	6320	0.99	6512	0.93
Japan	91760	0.75	92740	0.74
Norway	3526	0.9	3557	0.89
Denmark	4515	0.92	4520	0.93
Canada	17048	0.79	17442	0.80

This table is compiled from statistics given in the W.H.O. Epidemiological and Vital Statistics Report, Geneva, *1960*.

(178)

in Mexico. In 1951 the incidence of typhus and other ricketsial diseases in Ruanda Urundi was 135, of whom two proved fatal.

REFERENCES

1. WATT, A., 1959. *W. Afr. med. J.* **8**, *53*.
2. ONABAMIRO, J. S., 1949. Why Our Children Die. Methuen, London.
3. LAWSON, J. & LISTER, U., 1956. Report on the Work of Dept. of Obstetrics, University College, Ibadan.
4. BROADBENT, J & REIFF, B., 1956. *W. Afr. med. J.* **5**, *76*.
5. BODENHEIMER, F. S., 1951. Insects as Human Food. Dr. W. Junk Publishers, The Hague.
6. BASSIR, O., 1956. *W. Afr. med. J.* **5**, *28*.

CHAPTER II

DEVELOPMENT OF THE FOETUS

Haemoglobin

In man as well as other mammals, foetal haemoglobin differs from that of the adult. ANDERSCH and his colleagues[1] and BEAVAN et al.[2] have expressed this difference in terms of electrophoretic mobility and sedimentation constants. Derivatives of foetal haemoglobin have different solubility properties from those of adult[3]. Characterization of the free amino groups of the two types of haemoglobin attributed to the foetal haemoglobin a different structure from that of the adult[4].

In the earliest stages of pregnancy the blood of the foetus contains practically no adult haemoglobin, but the ratio of foetal to adult haemoglobin decreases during the course of gestation. TURNBULL & WALKER[5] showed that the concentration of foetal haemoglobin after 8 months pregnancy was 90% and that the ratio of foetal: adult haemoglobin had decreased to 4 : 1 by the end of pregnancy. This diminution in foetal haemoglobin continues after birth, reaching 50% at the age of four weeks[6], disappearing completely in full-term babies after 22 weeks[7].

The rate of disappearance of foetal haemoglobin from the blood of premature infants is slower. Various explanations have been put forward to account for this. The most feasible being (a) that normal erythropoiesis is slowed down in premature infants and (b) that erythrocytes lodging foetal haemoglobin continue to be produced in premature babies after they have been born.

If, as appears likely, the relative amounts of foetal and adult haemoglobin during pregnancy are concurrent with various stages in the nine-months cycle, then it is not surprising that at birth, and for weeks afterwards, the ratio of foetal: adult haemoglobin in the pre-mature infant should be higher than in the full-term baby.

WELBOURN[8] estimated the total haemoglobin of 8-weeks old East African children some of whose mothers were anaemic. He obtained no significant lowering of concentration, as compared with normal European children of comparable age. But the accuracy of his technique may have been open to question. He did not fractionate his haemoglobin into foetal and adult. Similar results were obtained by JORDAN[9].

It is interesting to note that the concentration of foetal haemoglobin in premature new-born infants is significantly higher than that of normal full-term babies. ABRAHAMOV et al.[10] obtained an

(180)

average foetal: adult haemoglobin ratio of 85.5% for the former and 65.1% for the latter.

This suggests that the concentration of foetal haemoglobin of the foetus' blood is physiologically linked with intra-uterine life, and most probably with respiration.

If oxygen were to be easily transferred from the maternal blood to the infant through the placenta, a pre-requisite would be that the dissociation curve, at the pH concerned, would be shaped in such a way as to give a lower percentage of oxygen saturation for the mother than the foetus at any given oxygen pressure.

EASTMAN, GEILING & DELAWDER'S experiments[11] suggest that such a mechanism is in fact at work in the latest stages of gestation. It appears that below 25 mm oxygen pressure the oxygen dissociation curve of human foetal blood is considerably steeper than that of the mother. The gap between the curves is widest at high oxygen pressure.

BARCROFT et al.[12] showed a progressive displacement of the oxygen dissociation curve of goat foetal blood towards that of the adult, until the two lines virtually overlap at parturition.

It is difficult to understand why solutions of foetal haemoglobin give oxygen dissociation curves which are situated to the right of the corresponding adult curves. And it is still inexplicable how the concentration of foetal haemoglobin is biologically controlled.

The evidence relating carbon dioxide concentration with the concentration of foetal haemoglobin is inconclusive. However, it is worth noting that LIEBSON et al.[13] found no difference in CO_2 dissociation curves, between foetus and mother, in a considerable series of experiments.

The enzyme, carbonic anhydrase, which controls the rate of the reaction

$$H_2 CO_3 \rightleftharpoons CO_2 + H_2O$$

has been shown to occur in the blood of the foetus and the new-born at a lower concentration than in the adult [14, 15]. The significance of this deficiency is not clear, since it has not been shown that the amount of this enzyme is related to anoxia in the human.

However, it is worthwhile to remember that LEWIS & ALTSCHULE[16] observed that in a form of anaemia which is related to a lowering of the concentration of carbonic anhydrase in the blood, exercise produced a greater respiratory distress than could be accounted for on the basis of the anaemia alone.

The connection between this observation and the anaemia of pregnancy to which tropical women are prone springs readily to mind and a study of the biochemistry of the blood of foetuses of malnourished and anaemic women, in connection with this enzyme, may prove profitable.

The physico-chemical properties of haemoglobin in foetal blood and that of infants have been investigated in tropical people by many workers, including HUISMAN et al.[17], SCHELL & McGINLEY[18]. Similar electrophoretic studies in other parts of the world have shown that different components are present in foetal and adult haemoglobin[19, 20, 21]. MOTULSKY and his colleagues have shown that on paper haemoglobin C has the least mobility towards the anodes. Adult haemoglobin moves fastest, while the sickle cell haemoglobin rate of migration lies between the two[20].

Sickle-cell-haemoglobin C appears to be restricted to Negro blood and causes a joint and bone disease which is most easily diagnosed by electrophoresis[18]. Up to 7% foetal haemoglobin (F) has been detected[22] in the blood of children with haemoglobin C. The sickle-cell haemoglobin of a Negro child of 7 years, studied by SCHELL & McGINLEY[18], had 54.7% haemoglobin C and 43.3% haemoglobin S. Her father had 23.8% haemoglobin C and her mother 28.5% haemoglobin S.

HENDRICKSE et al.[110] have reported the occurrence of a fast-moving haemoglobin, in 11 out of 100 new born Nigerian infants studied, which persisted for the first two months of life. It appears that both the "Bart's" and "Fessas-Papaspyron" types of haemoglobin were to be found among such cases. Sickling was detected at birth in only 5 out of the 100 babies examined, the percentage of foetal Hb being 50, 70, 89, 80 and 64 respectively. After two months of observation, six more sicklers were detected with foetal Hb concentration of 72% to 86%. Only one child developed the sickle-cell haemoglobin C disease.

SHOOTER, SKINNER and others[102] have undertaken the electrophoretic characterization of haemoglobin G using paper and moving-boundary techniques. In this way complete resolution of Hb G from haemoglobin A S & E have been obtained. Hb G_2 has been obtained by starch gel electrophoresis from blood specimens of patients suffering from haemoglobin G disease. Recently, BEAVEN et al.[103] have applied the above techniques and others for the identification and estimation of foetal haemoglobin.

According to GARLICK[104] young children in Western Nigeria whose blood contained sickling Hb were less prone to Malaria and they showed reduced rates and densities for *Plasmodium falciparum*. The distribution of Hb S between mothers and children suggested that pregnancy was relatively more successful in sicklers than in non-sicklers.

Although malaria is not an important factor determining the high incidence (27—37%) of haptoglobulin negatives in Northern Nigerians, since there is no difference in the distribution in young children and adults, environmental influences seem to be involved[105]. SMITHIES & WALKER[106] had shown earlier that three phenotypes 1—1, 2—1 and 2—2 can be detected by starch gel electrophoresis of serum proteins, in the α_2 globulin fraction which binds haemoglobin. Later ALLISON[107] described a modification of type 2—1 which is also stainable with benzidine in the starch gel. This phenotype is fairly common in American Negroes[108] and in Africans[105]. Transferrin, the iron-binding fraction of beta globulin of serum can be detected as several bands in starch gel electrophoresis, of which one band TfC is most constantly found. A slow moving TfD$_1$ has been discovered in American Negroes, Australian Aborigines and Nigerians[105, 109]. TfD$_1$ is determined by a single gene which is an allele of that controlling TfC.

In view of the relationship between iron-deficiency anaemia and protein malnutrition, further studies of transferrin reactions may indicate a link between long-term dietary deficiencies of those elements and the evolution of physiologically protective adaptations.

Mineral Metabolism

In the development of the foetus during gestation the main elements accumulated are calcium, phosphorus and iron. The two former are concentrated in the skeletal system while the latter is almost entirely restricted to the blood system.

As far as calcium is concerned, there is a gradual accumulation at the beginning of pregnancy followed by a steep rise in the last few weeks of gestation. Table V summarizes this change. Birth, whether premature or full term, reverses temporarily the rate of calcium retention as a percentage of body weight, even though the calcium content of the premature baby may be only 80% of the full-term value[23]. The effect of malnutrition of the mother upon the retention of calcium by the foetus has been studied by Genevieve

Table V

After SMITH[6]

Period of Gestation	Total fetal calcium	Increment during 2 months
Before 4 lunar months	1 g	
4 to 6 lunar months	3.7 g	2.7 g
6 to 8 lunar months	8.9 g	5.2 g
8 to 10 lunar months	21.0 g	12.1 g

STEARNS[23] and other workers[24, 25]. It would appear that minor dietary deprivations of the mother during pregnancy are not in general reflected in the rate of calcium retention in the foetus. But there is evidence that gross mineral malnutrition with severe hypo-calcaemia does affect the mineralization of the infant, although BOOHER & HANSMANN[28] were unable to find statistical difference in the mineralization of bone in children whose mothers had widely varying dietary intakes of calcium. TOVERUD[26] observed a high incidence of soft-skull and early rickets in children born by women who were undernourished with respect to calcium during pregnancy. Prematurity was also attributed to malnutrition in pregnancy. Using roentgenography, SONTAG[27] demonstrated poor bone struc-ture in such infants.

The experiments of FINOLA and his colleagues[25] showed that during pregnancy supplementation of the mothers' diet with larger amounts of vitamin D and calcium than the usual clinical doses may result in such increased calcification of the foetus, as to make labour distressing.

NICHOLLS[29] reckons that the calcium content of a newborn 7 lb baby in the tropics is less than 20 g and that the placenta holds less than 10 g. Hence he calculates an additional intake of 0.2 g calcium daily for a pregnant woman, making a total of 0.8 g per day. On this basis, the dietary intake of calcium of most tropical women is deficient during pregnancy[30]. It is therefore not surprising that many years ago, WRIGHT[31] described a 50% (of live births) inci-dence of congenital rickets in children born in Freetown, West Africa as follows:

"The posterior fontanelle was often very patent and connected with the anterior fontanelle by a very open sagittal suture some-times as much as half an inch wide, often a quarter of an inch. The two halves of the frontal bone were frequently found to be quite un-united and only joined by membrane bridging a gap an eighth of an inch or more. This condition is well recognised locally and is called 'occa'."

JELLIFE[32] and other observers[33, 34] have confirmed these early findings. However, apart from designating the name "pre-rachitic osteoporosis" no intra-uterine metabolic evidence has been forth-coming.

Since the daily breast milk output may contain up to 100 mg calcium, the requirement of this element during lactation would seem to be higher than in pregnancy. Therefore, the observations of HOLEMANS, LAMBRECHTS et al.[112] with lactating African mothers require confirmation, since they suggest that it is not necessary to supplement the diet of poorly nourished women with calcium during pregnancy. The daily mean calcium intake of the Congolese lac-tating mothers studied was 440 mg during the experimental period, and the average negative balance amounted to 47.5 mg. Yet, no adverse clinical signs were reported.

The metabolism of phosphorus is evidently related to that of calcium in the developing foetus, and nutritional provisions of the one element must depend on the availability of corresponding amounts of the other. TODD and his colleagues[35] reviewed twenty separate papers on this subject. When the concentrations of calcium and phosphorus in the blood serum of infants at birth are compared with those of the mother, calcium values ranging from 8.3 to 14.4 mg/100 ml were obtained for the new-born. The corresponding phosphorus values were 1.9—13.3 mg/100 ml[35]. The works of FINOLA et al.[25], HESS et al.[36], BOGERT et al.[37] and KRANE[38] put the average calcium concentration of maternal blood, at parturition, at approximately 1 mg/100 ml lower than the value for the new-born. A similar discrepancy occurs between the phosphorus concentration of serum of the mother and her child at birth.

However, the studies of BULLOCK[39] and TODD et al.[35] suggest that, whereas there is a decline in calcium retention in the infant after birth, there occurs a rise in the rate of accumulation of phosphorus. In this way, the product of Ca × P remains safely removed from that characteristic of tetany.

Of course, the body of the foetus of a malnourished mother contains less total phosphorus than the normal[40], but most of the phosphorus exists in bound, organic forms in the nucleic acids and nucleoproteins of the cell. The blood values quoted above are for the inorganic phosphorus. Similarly, the calcium of blood is partitioned into an ionized fraction which is hardly changed during the first few weeks of life, and a moiety bound to protein which alters[41, 42]. The measurement of total calcium (diffusable and non-filterable) is technically easy and reproducable, and it is in this way that TODD's values were obtained.

The enzymatic relationship between phosphatase, calcium and phosphorus is complex. For whereas the synthesis and general metabolism of bone cells from serum calcium and phosphorus is probably controlled by phosphatase, it is not certain how the activity or concentration of the enzyme is regulated.

SYDOW[43] found serum phosphatase activity of the prematurely born baby to be 50—100% higher than in full-term infants. He obtained an increase from 7.5 Bodansky units at birth to 17.5 units in the third week. BARNES & MUNKS's[44] results gave 7.1 Bodansky units as the average at birth, rising to 10.4 in the third week and only 11.4 units at the end of six months.

The possibility of easy irradiation of the skin by sunlight, in tropical countries, makes it unlikely that rickets can ever become an endemic disease. And it also means that vitamin D metabolism in the pregnant mother and her child is an important consideration in the development of the foetus.

A number of Chinese workers, including Woo et al.[45] and CHU & SUNG[46] have described incidences of tetany in the new-born which were probably due to low vitamin D intake of the mother during pregnancy. Foetal rickets has been studied in the Far East by other workers[47, 48].

JELLIFFE[49] lays much stress on the seclusion of Eastern women in Purdah in his consideration of the genesis of foetal rickets. If PLATT's hypothesis, quoted by JELLIFFE[49], of provitamin D activation of the skin is valid, the relationship between the vitamin A intake and the application of oils on the skin of pregnant women needs to be given special attention.

The mechanism by which iron is transferred from the blood of the mother to its foetus is not clear. Nevertheless, it would appear that trans-placental transfer of the metal does not involve prior synthesis of maternal erythrocyte. POMMERENKE's et al.[50] experiments with radioactive iron indicated that the element could be transferred across the placenta, to a foetus about to be born, within 40 minutes.

It has been shown that there is a steady increase in the serum iron content of foetus from an average value of 20 mg/100 ml, in the fifth month of pregnancy, to a mean of 150 mg/100 ml at birth[51, 52].

The work of LAURELL[53] suggests that during pregnancy a metabolic state of equilibrium is reached in favour of increased synthesis of haemoglobin from iron stores in the woman's body. For, whereas the iron binding capacity of maternal blood serum ranges from 300—450 mg/100 ml, the actual amount in transport in the serum decreases from 100 to 80 mg/100 ml during gestation. On the other hand, the foetus, at term, has an iron-binding capacity of only 225 mg/100 ml and a high serum iron value of 150 mg/100 ml.

It seems reasonable to conclude, therefore, that during pregnancy the maternal stores are depleted by increased formation of haemoglobin which is transferred across the placenta into the blood stream of the developing foetus. Over and above this, there is possibly another mechanism by which diffusible iron flows directly from mother to foetus. Whether the iron thus transferred is in ionic form or linked with amino-acids and peptides, is a matter of conjecture.

It is generally agreed that not all iron passed to the foetus from the mother is used in the formation of circulating haemoglobin. Assuming that the volume of blood of the newborn is 350 ml and that the concentration of total haemoglobin is then 17.5 g/100 ml, the total amount of circulating haemoglobin becomes 61.2 g. If the iron content of the haemoglobin is taken to be 0.336%, then the total blood haemoglobin iron is 0.336×612 mg i.e. 201.6 mg. SHOHL[54] reckons that the total body iron content of a new-born is about 375 mg. Therefore, we are forced to conclude that about 173 mg iron is stored in the liver and other tissues.

The range of concentration of liver iron obtained from the analysis of new-born tissue is 35—70 mg[6]. From all accounts, most of this is laid down during the last three months of pregnancy; but since the liver itself is growing rapidly then, the iron concentration per unit weight of the fat-free dry liver remains fairly constant throughout gestation[55]. How the remaining 100 mg of stored iron is distributed in the body of the foetus at full term is not known. The physiological significance of the stored iron in foetus tissue is probably inexplicable except in terms of the "physiological anaemia" which normally follows birth. Balance experiments conducted by STEARNS & McKINLEY[56] showed that there is a negative iron balance of 1—3 mg daily in the first month of neonatal life, if the baby is fed on evaporated milk. Since active haemoglobin synthesis commences in the third month, this early physiological anaemia is of little harm to the normal child.

However, the size of the iron store in the body of the new-born depends upon the intake of iron by the mother during gestation. GILLMAN & GILLMAN[57] found that the liver iron content of newborn infants of iron-deficient African mothers was less than a tenth of that found in well-nourished "European" fully developed foetus.

Although the haemoglobin content of the blood of newborn babies of malnourished women may not show a marked deviation from normal, the effect of poor intra-uterine iron storage is likely to be severely felt in neonatal life[58].

McLEAN[58] calculates that about 45 mg of iron may be added to the bodily stores of an infant, in the process of being born, by massaging the umbilical cord and draining it of blood until it ceases to pulsate, before it is severed.

When the maternal stores of iron are themselves low, such a treatment would tend to worsen the state of anaemia in which so many pregnant women in the Tropics enter labour; but there is usually a spontaneous remission of the anaemia after parturition. WOODRUFF[59] has correlated the anaemia of pregnancy of Yoruba women in Nigeria with their dietary intakes of protein and the metabolic functions of their liver, and he obtained an inverse proportion between the severity of the disease, as evinced by high incidence of prematurity and large erythrocyte diameter, and the amount of dietary protein.

The mean values for thymol turbidity and gold colloidal flocculation tests in a series of pregnant Nigerian women were five times higher when anaemia was present than when it was not, and the Bromsulphthalein retention and the alkaline phosphatase results were also higher. But EDOZIEN, ADENIYI-JONES & WATSON-

WILLIAMS[113] do not support the view that anaemia of pregnancy in Nigerian women is due to deficiency of protein or to liver disease.

Fatty Liver

The available clues as to the manner and extent of metabolic derangement in the foetus of malnourished persons are scanty and indirect. Yet, the observation of SILVERA & JELLIFFE[60] and EDINGTON[61], that fatty infiltration of the liver sometimes occurs before birth, suggests that liver dysfunction in intra-uterine life may be a direct sequel of malnutrition in the pregnant woman. It would appear that under normal conditions of pregnancy, foetal liver is unable to assume all the functions of an adult organ, even at full term.

Experimental work on animals[62, 63] has shown that deficiency of amino-acids or proteins may give rise to a lowering of the enzymatic content of the cell. This is a logical observation since enzymes are proteins, and these are in a constant state of "flux" and "turn over" in the living body.

The placental blood flowing into a foetus from the circulatory system of a woman, who is undernourished in respect to proteins, is likely (though not certain) to be deficient in some plasma proteins and their derivatives. In this case, we should expect a derangement of the mechanism of metabolic development in the foetus.

One way of assessing this kind of damage is that followed by WATERLOW[64] who estimated the contents of the enzymes serum cholinesterase, liver cholinesterase, liver cytochrome oxidase, liver lactic dehydrogenase and liver transaminase in very young Jamaican and Gambian babies and found a correlation between the nutritional state of the baby, the presence of histologically-detected fatty liver and reduced liver cholinesterase. On treatment of the subjects with proteins and other dietary factors, the concentration of the enzyme rapidly returned to normal. The concentrations of the enzymes in the liver were roughly paralleled by those of plasma. WATERLOW[64] failed to find any significant difference between the normal values and those of malnourished infants for all the other enzymes studied. The significance of this result and of later work[111] may be that homoeostatic control by the body is exercised to ensure that the level of these enzymes, whose roles are vital in general metabolism, is always maintained. It could also be that, as the cellular amounts of enzymes are not themselves important[65], their total turn-over rates would be a more delicate pointer.

Nevertheless, should cholinesterase come to be regarded as an important factor in the prevention of fatty liver in infants, and were the same pattern to be followed in the foetus, an explanation would be found for the reduction in the efficiency of the liver which is made manifest in the congenital fatty liver cases referred to above.

The fact that hyperbilirubinemia of the newborn is more pronounced in premature babies and steadily increases during the first week of life of full-term infants[66] suggests that the function of the liver, as far as the excretion of bilirubin is concerned, is never fully developed at birth.

The increased serum bilirubin has been shown to be due mainly to indirect-Van den Bergh-reacting pigments, and therefore not to stasis in the liver [67, 68]. Nor is it likely to be entirely dependent on the degree of haemolysis which is consequent upon the reduction at birth of the polycythenia of pregnancy anoxia[69].

SMITH[6] has reviewed the evidence obtained from testing the liver function of the new-born, including the bromsulphthalein excretion experiments of YUDKIN and his colleagues[70], and concluded that efficient liver function is not attained until about four days after birth.

If the liver of the foetus remained rudimentary until the last stages of pregnancy, and the organism needed the iron brought in the maternal erythrocytes, break-down of haemoglobin would be rapid and the excessive pigment would find its way into the umbilical artery, thereby causing a bilirubin gradient between foetus and mother. Such an observation has been made by CSERNA & LIEBMANN[71] and by FINDLAY[69]. However, the bloods of cord and of the new-born infant usually contain less bilirubin than that of a jaundiced mother[6]. This indicates a restriction to easy flow of bilirubin from mother to foetus.

General agreement has not been reached on BOOK's observation that the incidence of neonatal jaundice is higher in children who have received extra blood at birth via the umbilical cord[72], and it seems likely that the rise in blood bilirubin which occurs in the first few days of life in such cases "is present in every infant at birth and rises to its peak on the second to fourth day"[6].

Obviously, in the face of such universal hepatic immaturity, deficiency in the nutrition of the mother could adversely affect the function of the foetus' liver. The extent to which this kind of intra-uterine liver damage can affect the health of the individual in later life has not yet been established.

GOLDWATER & STETTEN[73] have shown that the foetal rat, at the latest stages of gestation, is utilizing 68% of its daily output of synthetic glycogen, and that the latter amounts to what its total body content was a few days earlier.

We know that in humans the store of glycogen in the placenta decreases and the glycogen content of foetal liver increases, as term is approached[74, 75]. But little is known of the metabolism of sugar and glycogen as sources of energy in intermediate metabolism in the foetus.

SHELLEY[114] suggests that foetal muscle glycogen is a source of carbon, hydrogen and oxygen required for the synthesis of tissue protein, and not of aerobic energy. HERTIG [115] has described the presence of glycogen in the human embryo germinal disc as early as the 13th day of pregnancy. A slight indication of how glycogen is formed in mammalian placenta is given by the application of the EMBDEN-MEYERHOF-CORI cycle [116]. Traces of fructose are present with glycogen in the human foetus and they probably serve the same function in the last days "in utero".

Evidence gained in short-term animal experiments with high-carbohydrate-low protein diets[76] show that the retention of glycogen by the liver is one of the earliest consequences of malnutrition in the young. If it is possible to apply such findings to the development of the human foetus, some light might be shed upon the genesis of congenital fatty liver disease.

Hormones

The changes which occur in the endocrinological make-up of a woman during pregnancy have been extensively studied and need not be laboured further. But the effects of these changes upon the foetus which dwells in this hormonal whirlpool for nine months, though obviously considerable, have not been completely inter-related biochemically.

DEANESLY[116] has recently summarized the present state of knowledge with respect to foetal endocrinology. It seems clear that physiological castration of a young foetus may affect the accessory sex organs in later life.

Much is owed to GILLMAN & GILLMAN[57] and TROWELL et al.[77] for their observations on the feminization of African male adults. The association of this phenomenon and liver disease has been studied for many years; the syndromes of gynaecomastia, testicular atrophy and loss of libido, decrease in axillary, pubic and body hair in African men have been noted by DAVIES[78]. In the female, dys-menorrhoea, menorrhagia, and uterine sub-involution are some of the affects of excessive Oestrogens.

TROWELL, DAVIES & DEAN[77] report that the male syndromes are frequently seen in Uganda men, and that the worst cases are usually found in those persons who show evidence of severe protein mal-nutrition. They further assert that sperm counts of men proved low in these circumstances. But they present little quantitative data, nor do they relate their results with fertility.

TROWELL[79] estimated an incidence of gynaecomastia of 5% among African railway porters in Nairobi. GILLMAN & GILLMAN[57] described clinical observations which are beyond the scope of this work.

(190)

RUSSELL & HARRISON[80] found a reduced sperm count in sub-fertile men but did not correlate these results with their state of nutrition. Recently, TEITELBAUM & GAULT[81] reported that the earliest stages of starvation in the dog were accompanied by increased spermatozoa counts. These authors wondered whether, if the same phenomenon were observed in man, it might not be a compensatory mechanism for maintaining species homoeostasis. Has this any relevance to the high fertility of Asians?

The issue which next faces us is whether the general tendency to feminization of malnourished adult male persons in the tropics has any connection with hormonal stimulus in intra-uterine life.

It has been shown that sex reversal may occur in some animal species by administration of large quantities of hormones to the mother in the early days of gestation[82]. Those female foetuses which received large amounts of oestrogen in this way were marked for life by having malformed genitals (which did not undergo the normal postnatal involution) and retarded growth.

The implication here is that the young infants' body has a limited power for countering the metabolic effects of maternal oestrogens. If we were to consider seriously the androgenic properties of 17-Ketosteroids, the observation of BURROWS et al.[84] that 17-Keto-steroid excretion in human pregnancy reached its peak at about the third month and was higher in women bearing male children than those who gave birth to female, may be significant. In man, androgens are said to be less concentrated in the placenta of a female foetus than in male[83].

The influence of maternal hormone on the foetus seems to be greater than that exerted through its own endocrinological apparatus because both male and female infants have excessive amounts of oestrogens in their blood[85, 86] representing about 70% of the total oestrogens circulating in the mother's blood[87]. Normally, the new-born excretes large amounts of oestrogens but the rate decreases to a low value by the end of the first week[88]. The excretion of pro-lactin and the secretion of "witches" milk are also associated with early infancy, irrespective of sex[94].

It is still not certain how maternal hormones are transferred across the placenta to the foetus and to what extent foetal oestrogen contributes to the high body content of the hormone at birth. What is even more important to this study is whether the concentration of female hormones in the new-born and their excretion in the urine is related to climate and race, or correlated with protein malnutrition. No answers to these questions have so far been reported.

If the foetal life is already conditioned to a maximum degree of growth and therefore unresponsive to added stimuli[16], then it is probably to the derangement of maternal hormone metabolism that we must turn for an explanation of the high adult femininity of tropical people.

It is highly speculative to suggest that protein malnutrition in the mother may lead to a diminishing donation of those factors which control the effect of female hormones on secondary sexual characteristics. This would account for the fact that sexual involution follows the characteristic pattern in the infancy of tropical peoples, while femininization increases in later life.

This hypothesis is subject to the inter-relationship of the hormones of the pituitary and the reproductive organs, and the mediating effects of the adrenal hormones. While considerable observations have been made by GILLMAN & GILLMAN[57] and by VINT[89] on the histological changes which occur in the pituitary gland of both normal Africans and those with cirrhosis or hepatic carcinoma, there are practically no records of the biochemical implication of such changes and foetal malnutrition.

It is interesting to note, however, that in the majority of the cases of chronic malnutrition studied, the GILLMANS found pituitary cysts. The increase in basophil cells of the pars posterior resembled that of "Cushing's syndrome in castrated persons". Apparently, the basophil count of the normal South African male is almost identical with that of the average European woman. But as the African figures are reported as increasing in severe liver damage[89], it is difficult to reach any conclusion as to the exact nutritional meaning of these histological findings.

HELLER & ZAIMIS[90] found that the concentration of posterior pituitary hormones (antidiuretic and oxytocic activity/mg dry neurohypophysial tissue) in newborn English infants was only a fifth of the amount found in adults. They related their findings to the fact that newborn infants appear unable to concentrate their urines to the same degree as adult persons.

Other workers have studied the function and development of the hypophysis in the foetus[92, 93]. It seems likely that pituitary function in intra-uterine life may become defective if there is an enzymatic derangement in the synthesis of "compound F" in the adrenal gland[95]. This compound has the vital role of inhibiting the secretion of adrenal androgens which is initiated by pituitary A.C.T.H.

Prior to the end of the 4th month of pregnancy, this enzymatic disorder will probably lead to female pseudohermaphrodism. Whereas if the onset is after sexual differentiation, only virilization will result[95]. It is tempting to think that the reverse sequence is possible and that this is related to the low protein intake of pregnant women in the tropics. After all, GILLMAN & GILLMAN[57] frequently observed gross adrenal atrophy and other structural changes, including poor differentiation of the cortical tissue, in poorly paid South Africans. And McCARRISON[97] showed, long ago, that chronic malnutrition is inimical to the adrenal gland. Although the results of BARNICOT's experiments[91] are awaiting full confirmation, it is not completely inexplicable that the urinary 17-Ketosteroid excretion of his African adult subjects is only about a half of the average values for Europeans.

In the absence of biochemical data, the danger of attributing the allegedly common incidence of hypothyroidism of Africans[98] to racial causes is real, especially in view of the fact that the basal metabolic rate may be lowered in a mild degree of malnutrition[77].

Cretinism is not of frequent occurrence among Africans although simple goitre is common in regions where the iodine intake is low. In these circumstances, the size, development and function of the foetal thyroid may be affected[99]. Work with radioactive iodine[100] leads one to conclude that iodine administered to pregnant women is increasingly bound in the foetal thyroid after the 14th week of gestation, and that the foetal thyroid is developed and controlling growth and metabolic processes for a considerable time before birth[101]. Thyroxin has been localized in the foetus three months after pregnancy.

Unfortunately, no record seems available on experiments designed to relate the iodine-binding activity of foetal tissue with deficiency of protein intake of the pregnant woman. It is not beyond the bounds of reason that severe protein malnutrition of maternal origin may adversely affect the function, size and development of the foetal thyroid, just as iodine deficiency does.

REFERENCES

1. ANDERSCH, M. A., WILSON, D. & MENTEN, M., 1944. *J. biol. Chem.* **153**, *301.*
2. BEAVEN, G. H., HOCH, H. & HOLIDAY, E. R., 1951. *Biochem. J.* **49**, *374.*
3. WYMAN, J., RAFFERTY, J. & INGALL, E., 1944. *J. biol. Chem.* **153**, *275.*
4. PORTER, R. R. & SANGER, G., 1948. *Biochem. J.* **42**, *287.*
5. TURNBULL, E. P. & WALKER, J., 1955. *Arch. Dis. Childh.* **30**, *111.*
6. SMITH, C. A., 1951. The Physiology of the New-born Infant. Charles C. Thomas, Springfield, Ill., p. 51.
7. JONXIS, J. H. P. 1949. in Haemoglobin – a Symposium. Interscience, New York.
8. WELBOURN, H. F., 1952. *E. Afr. med. J.* **29**, *131.*
9. JORDAN, P. 1954. *E. Afr. med. J.* **31**, *143.*
10. ABRAHAMOV, A., SALZBERGER, M. & BROMBERG, Y. 1956. *Amer. J. clin. Path.* **26**, *146.*
11. EASTMAN, N. J., GEILING, E. M. & DELAWDER, A. M., 1933. *Bull. Johns Hopk.* **53**, *246.*
12. BARCROFT, J., ELLIOTT, H. R., FLEXNER, L. B. et al., 1934. *J. Physiol.* **83**, *192.*
13. LIEBSON, R., LIKHNITZKY, I. & SAX, M., 1936. *J. Physiol.* **87**, *97.*
14. VAN GOOR, H., 1934. Doctorate Thesis, Groningen.
15. STEVENSON, S. S., 1943. *J. clin. Invest.* **22**, *403.*
16. LEWIS, H. D. & ALTSCHULE, M. D., 1949. *Blood* **4**, *442.*

32

17. HUISMAN, T. H. & PRINS, H. K., 1953. *J. Lab. clin. med.* **46**, *255*.
18. SCHELL, N. & McGINLEY, J., 1956. *Amer. J. Dis. Child.* **91**, *38*.
19. SMITH, E. W. & CONLEY, C. L., 1953. *Bull. Johns Hopk.* **93**, *94*.
20. MOTULSKY, A. G., PAUL, M. H. & DURRUM, E. L., 1954. *Blood*. **9**, *897*.
21. SPAET, T. H., 1953. *J. Lab. clin. med.* **41**, *161*.
22. SINGER, K., CHAPMAN, A. Z., GOLDBERG, S. R., RUBINSTEIN, H. M. & ROSEN-
    BLUM, S. A., 1954. *Blood* **9**, *1023*.
23. STEARNS, G., 1939. *Physiol. Rev.* **19**, *415*.
24. SWANSON, W. W. & IOB, V., 1939. *Amer. J. Obstet. Gynec.* **38**, *382*.
25. FINOLA, G. C., TRUMP, R. A. & GRIMSON, M., 1937. *Amer. J. Obstet. Gynec.*
    34, *955*.
26. TOVERUD, K. U., 1937. *Acta paediatr. Stockh.*, **22**, *91*.
27. SONTAG, L. W., 1938. *Amer. J. Dis. Child.* **55**, *1284*.
28. BOOHER, L. E. & HANSMAN, G. H., 1931. *J. biol. Chem.* **94**, *195*.
29. NICHOLLS, L., 1951. Tropical Nutrition. Bailliere, Tindall and Cox,
    London.
30. BASSIR, O., 1951. Doctorate thesis, London.
31. WRIGHT, E. J., 1926. *Ann. Med. San. Rep. Sierra Leone*.
32. JELLIFFE, D. B., 1951. *Trans. R. Soc. trop. Med. Hyg.* **45**, *143*.
33. MILLOT, J., 1941. *C. R. Acad. Sci.* **213**, *370*.
34. WRIGHT, E. J., 1951. *Trans. R. Soc. trop. Med. Hyg.* **46**, *104*.
35. TODD, W. R., CHUINARD, E. G. & WOOD, M. T., 1939. *Amer. J. Dis. Child.*
    57, *1278*.
36. HESS, A. P. & MATZNER, M. J., 1923. *Amer. J. Dis. Child.* **26**, *285*.
37. BOGERT, L. J. & PLASS, E. D., 1923. *J. biol. Chem.*, **56**, *297*.
38. KRANE, W., 1930. *Z. Geburtsh. Gynäk.* **97**, *22*.
39. BULLOCK, J. K., 1930. *Amer. J. Dis. Child.* **40**, *725*.
40. SONTAG, L. W., MUNSEN, P. & HUFF, E., 1936. *Amer. J. Dis. Child.* **51**, *302*.
41. ANDERSCH, M. & OBERST, P. W., 1936. *J. clin. Invest.* **15**, *131*.
42. RAY, H. H. & PHATAK, N. M., 1930. *Amer. J. Dis. Child.* **40**, *549*.
43. SYDOW, G., 1946. *Acta paediatr., Stockh.* **33**, Supp. 2.
44. BARNES, D. J. & MUNKS, B., 1940. *Proc. Soc. exp. Biol. Med.* **44**, *327*.
45. WOO, T. T., FAN, C. & CHU, F. T., 1941. *Chin. med. J.* **60**, *99*.
46. CHU, F. T. & SUNG, C., 1940. *J. Pediat.* **16**, *607*.
47. MAXWELL, J. P., 1935. *Proc. R. Soc. Med.* **28**, *265*.
48. MAXWELL, J. P., PI, H. T., LIN, H. A., & KUO, C. C. 1939. *Proc. R. Soc. Med.*
    32, *287*.
49. JELLIFFE, D. B., 1955. Infant Nutrition in the Sub-tropics and Tropics,
    W. H. O., Geneva. p. *94*.
50. POMMERENKE, W. T., HAHN, P. F., BALE, W. F. & BALFOUR, W. M., 1942.
    Amer. J. Physiol. **137**, *164*.
51. VAHLQUIST, B. C., 1941. *Acta paediatr., Stockh.* **28**, Supp. V.
52. BRENNER, W., 1948. *Z. Kinderheilk.* **65**, *727*.
53. LAURELL, C. B., 1947. *Acta physiol. scand.* **14**, Supp. 46.
54. SHOHL, A. T., 1939. Mineral Metabolism. Reinhold Publ. Co., New York.
55. IOB, V. & SWANSON, W. W., 1938. *J. biol. Chem.* **124**, *263*.
56. STEARNS, G. & McKINLEY, J. B., 1937. *J. Nutrit.* **13**, *143*.
57. GILLMAN, J. & GILLMAN, T., 1951. Perspectives in Human Malnutrition.
    Grune and Stratton, New York.
58. McLEAN, E. B., 1951. *Pediatrics* **7**, *136*.
59. WOODRUFF, A. W., 1951. *Brit. med. J.* **2**, *4745*.
60. SILVERA, W. D. & JELLIFFE, D. B., 1952. *J. trop. Med. Hyg.* **55**, *73*.
61. EDINGTON, G. M., 1954 in: Malnutrition in African Mothers, Young Children
    and Infants. H. M. S. O., London.
62. SCHULTZ, J., 1949. *J. biol. Chem.* **178**, *451*.
63. WILLIAMS, J. N., DENTON, A. E. & ELVEHJEM, C. A., 1949. *Proc. Soc. exp.*
    Biol. Med. **72**, *386*.

64. WATERLOW, J., 1954. in: Malnutrition in African Mothers, Infants and Young Children, H. M. S. O., London.
65. POTTER, V. R., 1949. in: Respiratory Enzymes. Burgess, Minneapolis.
66. DAVIDSON, L. T., MENITT, K. K. & WEECH, A. A., 1941. *Amer. J. Dis. Child.* **61**, *958.*
67. HIRSCH, ADA, 1913. *Z. Kinderheilk.* **9**, *196.*
68. LEPEHNE, G., 1922. *Mschr. Geburtsh. Gynäk.* **60**, *677,*
69. FINDLAY, L., 1947. *Arch. Dis. Childh.* **22**, *65.*
70. YUDKIN, S, GELLIS, S. S. & LAPPEN, F., 1949. *Arch. Dis. Childh.* **24**, *12.*
71. CSERNA, S. & LIEBMANN, S., 1923. *Klin. Wschr.* **2**, *2122.*
72. BOOK, N., 1935. *Canad. med. Ass. J.* **33**, *269.*
73. GOLDWATER, W. A. & STETTEN, DE W., 1947. J. *biol. Chem.* **169**, *723.*
74. CLOGNE, R., 1924. *Gyn. et Obstétrique* **10**, *23.*
75. TRAMA RAS, G., 1933. *Riv. Ital. ginec.* **15**, *351.*
76. BASSIR, O., 1955. *W. Afr. med. J.* **4**, *78.*
77. TROWELL, H. C., DAVIES, J. N. & DEAN, R. F., 1954. Kwashiorkor. Arnold, London.
78. DAVIES, J. N., 1949. *Brit. med. J.* **2**, *676.*
79. TROWELL, H. C., 1949. *E. Afr. med. J.* **25**, *236, 311, 423.*
80. RUSSELL, J. K. & HARRISON, R. G., 1953. Studies on Fertility. Charles C. Thomas, Springfield, Ill.
81. TEITELBAUM, H. & GAULT, W. H., 1956. *Science* **124**, No. 3217, *363.*
82. GREENE, R. R., BURRILL, W. M. & IVY, A. C., 1939. *Amer. J. Anat.* **65**, *415.*
83. CUNNINGHAM, B. & KUHN, H. H., 1941. *Proc. Soc. exp. Biol. Med.* **48**, *314.*
84. BURROWS, H., MACLEOD, D. H. & WARREN, F. L., 1942, *Nature, Lond.* **149**, *300.*
85. SEIGERT, F. & SCHMIDT-NEUMANN, 1930. *Zbl. Gynäk.* **54**, *1630.*
86. SOULE, S. D., 1938. *Amer. J. Obstet. Gynec.* **35**, *309.*
87. SKLOW, J., 1942. *Proc. Soc. exp. Biol. Med.* **49**, *607.*
88. PHILIPP, E., 1938. *Klin. Wschr.* **17**, *797.*
89. VINT, F. W., 1949. *E. Afr. med. J.* **26**, *58.*
90. HELLER, H. & ZAIMIS, J. E., 1949. J. *Physiol.* **109**, *162.*
91. BARNICOT, N. & WOLFFSON, D., 1952. *Lancet* **1**, *893.*
92. HALPERN, S. R., 1938. *Endocrinology* **22**, *173.*
93. CHANNICK, B. J. & SOKHOS, D., 1956. J. *Pediat.* **49**, *80.*
94. DOBSZAY, L., 1938. *Amer. J. Dis. Child.* **56**, *1280.*
95. KELLEY, V. C., ELY, E. S. & RAILE, R., 1953. *Paediatrics* **12**, *541.*
96. WILKINS, L., 1950. The Diagnosis and Treatment of Endocrine Disorders in Childhood and Adolescence. Charles C. Thomas, Springfield, Ill.
97. McCARRISON, R., 1919. *Indian J. med. Res.* **6**, *275.*
98. DIERCKZ, F. & HELMONS, J., 1950. *Ann. Soc. Belge Méd. trop.* **30**, *411.*
99. NEUMANN, H. H., 1937. *Arch. Gynäk.* **163**, *368.*
100. CHAPMAN, E., CORNER, G., ROBINSON, D. & EVANS, R., 1948. J. *clin. Endocrinol.* **8**, *717.*
101. COOPER, E. R. A., 1925. The Histology of the more Important Human Endocrine Organs at Various Ages. Oxford University Press, New York.
102. SHOOTER, E. M., SKINNER, E., GARLICK, J. P. & BARNICOT, N., 1960. *Brit. J. Haemat.* **6**, *140.*
103. BEAVEN, G. H., ELLIS, M. & WHITE, J. C., 1960. *Brit. J. Haemat.* **6**, *1.*
104. GARLICK, J. P., 1960. *Trans. R. Soc. trop. Med. Hyg.* **54**, *146.*
105. BARNICOT, N. A., GARLICK, J. P. & ROBERTS, D. P., 1960. *Ann. Hum. Gen.* **24**, *171.*
106. SMITHIES, O., & WALKER, N. F., 1956. *Nature, Lond.* **176**, *1265.*
107. ALLISON, A. C., 1959. *Nature, Lond.* **183**, *1312.*
108. GIBLETT, E. R., 1959. *Nature, Lond.* **183**, *192.*
109. GIBLETT, E. R., HICKMAN, C. & SMITHIES, O., 1959. *Nature, Lond.* **183**, *1589.*

34

110. HENDRICKSE, R. G., BOYO, A. E., FITZGERALD, P. A. & RANSOME KUTI, S., 1960. *Brit. med. J.* **1**, *611*.
111. WATERLOW, J., 1959. *Med. Prac.* **18**, *1143*.
112. HOLEMANS, K., LAMBRECHTS, A., HULEH, C. & MARTIN, H., 1959. *J. trop. Paediat.* **5**, *27*.
113. EDOZIEN, J. C., ADENIYI-JONES, C. & WATSON-WILLIAMS, E. J., 1961. *Brit. med. J.* **i**, *336*.
114. SHELLEY, H., 1960. *J. Physiol.* **153**, *527*.
115. HUGGETT, A. ST. G., 1961. *Brit. med. Bull.* **17**, *122*.
116. DEANESLY, R., 1961. *Brit. med. Bull.* **17**, *122*.

CHAPTER III

FEEDING AND GROWTH OF THE YOUNG INFANT

Lactation and Breast-feeding

The biochemical processes which lead to the onset of lactation are mainly hormonal in origin. The breasts of a pregnant woman enlarge under the stimulating influence of increased oestrogen secretion. Their secretory tissues develop and proliferate by the aid of progestins which are formed in the placenta and corpus luteum. Finally, mamatropin, the prolactin secreted by the pituitary for this purpose, induces the generation of milk.

Normally the dripping of breast milk commences a day or two after parturition and lasts for many months, depending on a large number of different factors which will be considered later. The young infant may be fed entirely on breast milk or on cows' milk and other formulae.

In the beginning the flow of the milk is slow, but it gradually increases until full lactation is achieved on about the fourth day[1]. Then the breasts feel warm and full and squirt milk freely when expressed or sucked.

This squirting is probably due to the action of oxytocin from the mother's pituitary gland; the secretion of the hormone is consequent upon the setting up of a nervous stimulus which initiates from the breast being sucked. It is said that oxytocin acts upon the myoepithelial contractile cells in the walls of the alveoli[1]. Apart from being largely responsible for the rhythmic production of the milk-ejection reflex in the mother, the baby further contributes to lactation by removing the secreted milk from the sinuses of the breast.

The milk which is produced during the first four to seven days of lactation is colostrum that is secreted in the alveolar cells of the gland from the seventh month of gestation until full term. The correlation between the change of colostrum to milk and the output of steroid hormones has been studied in Nigerian women[125].

Not only at this stage, but throughout the period of lactation, it would seem that the neuro-hormonal reflex produces a single stream of milk, and therefore through one breast at a time. The milk at the start of feeding from one breast usually contains a more uneven distribution of fat than that which is subsequently obtained from the other gland[1].

Although full lactation may not be obtained for as long as ten days after birth[1], insufficiency of milk may arise from other causes such as inadequate emptying of the breast, fatigue, poverty and malnutrition, genetic and other factors[2].

(197)

WALLER[3] made extensive observations on the connection between milk yield and the proportion of milk left behind in the breasts after sucking. He found that by expressing manually after feeding, as much as 150% of the volume of milk taken by the child could be obtained.

ILLINGWORTH[2] maintains that premature use of complementary feeds is the chief cause of the failure of lactation.

That psychological factors play their part in causing the failure of milk supply has been underlined by MOLONEY[4] who described how bombing stopped lactation in a number of Japanese women during the last war. NEWTON & NEWTON's[5] drastic experiments in order to demonstrate the central inhibition of the draught reflex are well known, and have been referred to elsewhere[6] in connection with failure of lactation in Nigerian townswomen.

When delivery takes place in hospital and the mother has to lie in for a fortnight, continuous interruptions by doctors, nurses, cleaners and laboratory staff could lead to physical exhaustion since these interviews and the feeding routine may extend over 18 hours daily. In the tropics, once delivery is completed and the mother has left her bed, she has to combine the strenuous physical activity of day-to-day living with that of tending to her young. "Such fatigue inevitably has a deleterious effect on lactation"[2].

It has still not been established beyond doubt that under-nutrition is an invariable cause of diminished lactation in man. It is logical to assume that, if the mother's diet is deficient, her breast milk would lack the corresponding ingredients. The results of the many workers who have expressed conflicting opinions on this matter are themselves quantitatively inadequate. Thus TOVERUD[7] equated the lengthening of the period of lactation with an increased output. He found that after supplementing the diets of working class women with vitamins and other nutrients the mothers were able to breastfeed their babies longer than otherwise. A similar criterion was used by DEAN[25], and by ABELS[8], ANTONOV[9] and SMITH[10], but these are all agreed that malnutrition and starvation do not influence lactation for the worse. Even the chemical analyses of WALKER et al.[11] and KROPMAN[12], who studied the composition of breast milk of malnourished Africans, are not conclusive because (a) no objective criteria of dietary intake or metabolic disorder were used in the assessment of the nutritional states of the subjects and (b) no information was obtained on the total 24-hour output of milk or any of the constituents considered. Nevertheless, WALKER and his colleagues maintain that either the role of diet in lactation is over-emphasised, or, alternatively, "the body has a greater capacity to adapt itself to an inferior diet than is usually believed."

While genetical and racial factors may play a major role in the production of milk insufficiency in man, little experimental evidence is available. Some of these have been discussed by BASSIR[6]. In quantitative studies involving some 200 Lagos women, it has not been possible to discover any characteristically Negroid feature of lactation[6, 13]. Agalactia is unknown in most parts of Africa.

Peasant women generally maintain lactation over a much longer period than sophisticated mothers[11, 14], but this is probably neuro-hormonal in basis and is caused by constant sucking at the breasts. There is ground for the belief that shortened period of breast-feeding of some tropical people is related, in some way, with increased sophistication[6]. Also, it is common knowledge that lactation can be stimulated and maintained, in wet nurses, over long periods even though they have not recently delivered babies themselves.

CANET[17] feels that the volume of milk produced by Indo-Chinese peasants becomes reduced in volume quite early in the period of lactation, although the flow may be maintained for over two years. The same pattern of prolonged lactation has been described in detail for China, India, Thailand, Africa, Peru, Indonesia[18] but not for the West Indies and Brazil[19, 20] MARGARET MEAD[21, 22], referring to the cultural contexts of nutritional patterns in New Guinea, described the popular belief that by drinking plenty of coco-nut water and having the breasts sucked constantly, women who have never borne children of their own can be induced to produce large quantities of milk sufficient to nurse a young infant.

Agidi-baba, a kind of maize-gruel, is attributed similar galacto-genic properties by certain Yoruba people in Nigeria. Herbal galac-togogues are of widespread use in many parts of the tropics. Belief in such means of maintaining lactation possibly stems from the fact that mothers often feel thirsty when the draught-reflex is operated. However, good quantitative evidence of the milk-producing effects of carbohydrate drinks does not seem to be available.

WOODRUFF[23] believes that enough water or other fluids should be taken by the nursing mother in the tropics to ensure a daily urinary output of two pints. It is uncertain as to how he arrived at this figure. The same worker recommends that about 15 grams of common salt should be incorporated daily into the food of lactating mothers in hot climates.

According to JELLIFFE[18] the concentration of the vitamins, thiamine, ascorbic acid, vitamin A and riboflavine is dependent upon the state of nutrition of the mother. He quotes the following as normal values for human milk:

Substance	Concentration
Protein	1%
Casein	0.45%
Albumin	0.44%
Fat	5%
Calcium	28 mg/100 ml
Phosphorus	14 mg/100 ml
Iron	0.1 mg/100 ml
Vitamin A	170-670 I U/100 g
Thiamine	9-15 mg/100 g
Riboflavine	28-62 mg/100 g
Niacin	66-330 mg/100 g
Vitamin C	2-6 mg/100 g

The detailed dietary surveys and careful chemical analysis of breast milk which were undertaken by KON and his co-workers in England[24] throw considerable doubt on the veracitv of the statement that the state of nutrition of the mother affects the vitamin content of the milk, for they found "very little correlation between the vitamin content of the milk and that of the diet." In general, the results of their analysis yielded values which were very close to the upper limits of JELLIFFE's results, in full lactation (assuming that the values tabulated above for thiamine, riboflavine, and niacin were meant to be in μg/100 g).

In an experiment designed to study the effect of nutrition on the composition of breast milk, DEAN[25] divided a group of newly-delivered German women into three groups (A— C), irrespective of whether they were grossly malnourished or not. Group A had fat supplements of about 150 g daily. Extra proteins and carbohydrates were given to the women in group B, increasing their daily calorie intake from 2200 to 3300. The mothers classified in group C were used as controls and had no supplements. The basic diet, which all the subjects had, consisted of 420 g carbohydrates, 76 g protein, and 34 g fat. Apart from a transcient increase in the milk fat in the evening, following the administration of extra lipid to the diet of the group A women, no obvious effect was seen in the composition of the breast milk of the experimental subjects. DEAN himself concluded that the intake of fat does not seem ordinarily to determine the percentage of fat found in milk, since the normal diurnal variation far exceeded any change that might have been attributed to the fat supplementation. Milk was expressed periodically from one breast while the other was being sucked.

The average daily output (ml) of milk of the three groups of experimental subjects on the 8th day were (A) 175.8, SD. = 47.0; (B) 226.2, SD = 79.0; (C) 257.8, SD = 81.0. These values represent well over 90% of the amount estimated, by test weighing, as being

Table VI

Composition of Lagos Mothers' Milk in Full Lactation*

	Cream %	Nitrogen %	Protein N × 6.25	Lactose %	Calcium mg/100 ml	Phosphorus mg/100 ml	Ca/P
Number of subjects	168	182	182	174	146	154	144
Range of results	1.8–16.0	0.03–1.17	0.15–7.3	3.15–10.4	10.0–32.0	4.0–17.7	1.02–3.76
Mean Values	5.87	0.264	1.628	7.24	20.3	9.54	2.18
Standard Deviation	2.54	0.12	0.75	1.25	4.68	2.22	0.44
Estimated mean for European & American milk (MORRISON[127])	Total lipoid 3.33 S.D. ±0.57		1.316 S.D. ±0.32	7.227 S.D. ±0.67	29.9±0.5 KON & MOWSON[24]	14.1±0.1 MACY[32]	

* After BASSIR[13]. Reproduced by courtesy of the publishers of the Journal of tropical Medicine and Hygiene.

(201)

sucked from the other breasts. No significant differences in milk proteins and fats were observed.

Moreover, the mean values for ascorbic acid, riboflavine, and total nitrogen of 60 of the German women in full lactation in 1946—47 (i.e. not long after the period of extreme food shortage) were approximately the same as those of KON & MAWSON[24]. Ascorbic acid concentration ranged from 3.2—4.3 mg/100 ml, riboflavin from 18.8 to 25.1 µg/100 ml and total nitrogen varied from 186 to 227 mg/100 ml. It is, therefore, surprising that TROWELL et al.[26] expressed the opinion that the ability to lactate could be affected "by even a moderate degree of undernutrition".

The mean 24 hour volumes obtained by DEAN[25] are very similar to those recorded by HYTTEN[27] with fairly well nourished Scottish women, and by BASSIR (Table VI)[13] with ordinary Nigerian townfolk.

AUFFRET & TANGUY's[28] analysis of the milk of poor African mothers in French West Africa yielded low values for some of the essential amino acids, especially methionine, but this observation has not been confirmed[29, 116]. In his estimate of the sulphur-containing amino acids of African breast milk, BIGWOOD[30] nevertheless assumed that there is no difference between the milks of European and African women. On this basis, it is possible to draw up a table comparing the amino acid constituents of cow and breast milk as shown in Table VII.

If the caloric requirement of a year old baby were taken to be 1000, an intake of 1500 ml milk would be obligatory, if the child was fed on breast milk alone[30]. The amino acid sulphur equivalent of this volume of milk works out at a little over 0.13 g. Should it be necessary for some of the dietary sulphur to be diverted, in cellular metabolism, to processes involved in the detoxication of food poisons, e.g., HCN from cassava, the actual amount of protein sulphur required will be larger.

The distribution of the major constituents of Nigerian mothers' milk during the first year of lactation has been statistically analysed by BASSIR[6, 13] and a comparison with European and North American mothers' milk is made in Table VI. SRINIVASAN & RAMANATHAN[33] have conducted similar studies on the breast milk of 49 Indian women of very low socio-economic standing with similar results as those shown in Table VI. GOPALAN[34] assessed the nitrogen intake of breast fed babies of 14 low caste women in the age range 20 to 30 years, by using an arbitrary average protein content of 1.2% which is very near to the figure of 1.19 which was obtained by SRINIVASAN & RAMANATHAN[33]. By allowing the babies to be fed on demand, and test-weighing them before and after each feed, GOPALAN determined the 24 hour output of each mother at weekly intervals for the first four weeks, and fortnightly for the next 20

Table VII

Comparison of the Content of Amino Acids of Bovine and Human Milk Proteins.
Amino Acid (g/16.0 g of Nitrogen)*

Constituent	Arginine	Histidine	Lysine	Tyrosine	Trypto-phane	Phenyl-alanine
Cow's Milk	4.3	2.6	7.5	5.3	1.6	5.7
Human Milk	4.3	2.8	7.2	5.2	1.9	5.6

Constituent	Cystine	Methio-nine	Threo-nine	Leucine	Isoleu-cine	Valine
Cow's Milk	1.0	3.4	4.5	11.3	8.5	8.4
Human Milk	3.4	2.2	4.6	9.8	7.5	8.8

* Based on a paper by BLOCK & MITCHELL [31]

Sulphur (g/100 g protein)+

Constituent	Methionine	Cystine	Total
Cow's Milk	0.452	0.214	0.666
Human Milk	0.365	0.40	0.765

+ Based on a paper by BIGWOOD [30].

weeks. The mean 24 hour volumes thus obtained lay in the surprisingly narrow range of 16 to 18.6 oz (353—527 g), while the average protein intake/kg body weight decreased from 2 g in the first fortnight to about one gram in the sixth month.

It is difficult to interpret the significance of GOPALAN's observations since no detail was given of the individual weekly variation in output, or of the standard deviations of the means. The value of the work is further diminished by the fact that (a) the diets of his very poor subjects were only vaguely assessed by an 'oral questionnaire technique', and (b) the possible variations of the protein contents of the milks was, at no time during the whole experiment, tested by chemical analysis.

As far as African milk is concerned, perhaps the most important deviations from the European and North American normal are the low values for calcium which are recorded in Table VI above, and the fact that despite the low absolute values, the Ca/P ratio is not significantly altered.

The effects of parity on these constituents and on the total output of milk have been investigated in African women[6] but the results are not clear-cut. It would seem, however, that primiparae are more consistent in their lactation than multiparae.

NICHOLLS[34] referred to the low dietary intakes of calcium and phosphorus by breast fed Ceylonese, and UGA[35] found the same low values for the contents of calcium and phosphorus in the milk of Japanese women. It is possible that man successfully adapts himself to a low intake of these minerals.

The biochemical importance of the migration of breast milk proteins in an electric field is a subject which is arousing great interest among research workers. It has been suggested[29] that the relative amounts of lactalbumin, lactoglobulin and caseins not only vary widely in the normal African, but that they may be associated with the protection of the baby against ailments such as malaria. The main difficulty in this kind of investigation is to establish, beyond doubt, what is normal and what is not.

When the requirement of all the dietary constituents of breast-milk is fully met, the growth rate of the infant may be considered optimal. As GOPALAN[34] has pointed out, the absolute weight of an infant may be so low, to start with, that it might be fallacious to use the rate of increase, with time, as an objective criterion of adequate nutrition.

In the first two years, any positive correlation between birth weight and birth rank is rapidly eliminated, provided the infant is provided with adequate nutrition throughout that period[36]. But it is well known[37, 38, 39] that climate, disease and other factors seriously hamper growth in tropical infants after the first six months, whether they are breast-fed or not. So it is to be expected that, in the Tropics, the averages of both the birth weight and the growth rate may be lower than those of North American or European babies (cf. [118]).

In GOPALAN's experiment[34] with the children of very poor Indians during the first six months the birth weights of the Indians were, on average, about a third less than the normal quoted for American infants; yet the growth curves appeared to run parallel. The protein intake of 2 g/kg body weight was halved in the Indians by the end of the experiment, when the growth rate was running parallel with the American curve[79]. The slope of the growth curve of 26 malnourished Gambian children studied by WALTERS & WATERLOW[79] was approximately half that of United States babies, as given by HOLT & MCINTOSH[80], between the ages of four months and two years. GOPALAN[34] discussed the protein requirement of infants, on the basis of an average protein concentration of 14 g per 100 g fat-free body of a new-born baby[40], and concludes that "the actual requirements are not more than 2 g per kg in the first few weeks, and probably even less in the later stages."

It is almost impossible, certainly unwise, to state a definite figure as the requirement for calcium. The rate of linear growth of a normal breast-fed baby is said to be greater than that of an artificially-fed baby having the same calcium retention[41]. Yet, linear growth of infants on cow's milk is proportional to the amount of calcium retained[42]. The average calcium content of an infant on breast milk decreases from 0.8% at birth to 0.55% at three months. With cow's milk, the concentration at three months of age is 0.7%. Thereafter, the growth curve has about the same slope as that of the foetus[43]. This supermineralization indicates storage of calcium, but it is uncertain that such storage is of any intrinsic metabolic value. It seems possible that the requirement of this element varies widely with individuals and climate. It would obviously be worthwhile to obtain analytical data on the total body calcium contents of Asian and African babies, in normal and abnormal growth, and to compare these with retention figures and protein intakes.

The amount of phosphorus in the body and the daily requirement of this element depend on the dietary protein and calcium. Hence, with human milk, the requirement is less than with cow's milk. More phosphorus is excreted in the urine of babies fed on cow's milk than in that of the breast-fed. This suggests that the amount of phosphorus taken in artificial feeding is not necessarily beneficial. It would appear that African children thrive well on a daily intake of 50—100 mg phosphorus in breast milk[6, 13].

The retention of calcium, and therefore of phosphorus, is enhanced by the amount of vitamin D present in the food. It is generally agreed that the vitamin D content of breastmilk is too low to ensure optimum growth and prevent rickets in infants[44, 45, 46] HENRY, KON & MAWSON[46] obtained seasonal variations ranging between 0.13 I.U./g fat in midwinter and 0.27 I.U./g fat in mid-summer. Those values are similar to those obtained in America by POLSKIN and his colleagues[44], but much lower than the results of DRUMMOND and others[47] who found no seasonal variation in the vitamin D concentration of the milk of English women.

The effect of irradiating women with a view to increasing the vitamin D content of their milk has been studied by HESS & WEINSTOCK[49] who were successful in enhancing the curative effect of breast milk on rat rickets by irradiating the mother. However, there are not sufficient quantitative data on the effect of irradiation and vitamin D secretion in breast milk to warrant any firm conclusions.

JELLIFFE[18] found that rickets developed in Nigerian babies whose mothers were over-dressed, but he presented no evidence that this was due to either (a) the under-production of vitamin D in the mother or the child, or (b) to the cumulative effect of both. It is just possible that because the diets of nursing mothers in some parts of the Tropics tends to be deficient in cholesterol and other lipids, 7-dehydrocholesterol synthesis in the skin may be retarded in such cases. This would result in reduced vitamin D production from solar irradiation.

It must not be forgotten, however, that it is not easy to increase the vitamin D content of breastmilk by giving the vitamin to the mother either in the state of pregnancy or during lactation. Administration of very large doses of vitamin D resulted in insignificant rise in milk vitamin.

JEANS[41] takes the view that no more than 350 I.U. vitamin D daily is the requirement for optimum growth if the vitamin is dispersed in lipid. It has also been shown that in fact the growth rate of infants may be retarded if the vitamin A intake is above 1800 I.U. per day[48].

The Food and Nutrition Board of the United States National Research Council recommends a daily intake of 1500 international units of vitamin A at the age of six months. This is equivalent to 500 µg of the vitamin[46]. And it is less than the supply of the vitamin in breast milk of well nourished mothers[51, 50]. The curve which best fitted the scatter diagram of HENRY et al.[46], showing the variation of vitamin A content of milk fat with the stage of lactation, suggests that at six months the total amount of the vitamin which is supplied in breast milk is about two-thirds of its value in the first month.

McCOSH, MACY and others[52], using a biological technique, found no increase in the vitamin A content of the milk of three women each of whom were given 15 g cod liver oil a day. The women were already on an adequate diet. WITH & FRIDERICHSEN[53] reported that it was not possible to increase the vitamin A content of breastmilk by the administration of large doses of β-carotene to the lactating mother. This observation was confirmed by HENRY, KON & MAWSON[46].

In an experiment with 47 samples of milk obtained from lactating mothers in Wuppertal who had been feeding on large amounts of vegetables, the vitamin A content averaged 39.7 I.U. per g fat with a standard deviation of 14.1. The observer, DEAN[25], reckoned that his figures were higher than those of HENRY et al.[46] and that this was probably due to the high carotene content of the vegetables consumed by the slightly undernourished German women. But the validity of DEAN's statistical assumptions (concerning the fall in vitamin A with the stage of lactation), which made the comparison possible, may not be absolute.

There does not appear to be any indisputable evidence as to the effect of giving β-carotene to a deficient mother on the vitamin A level of her milk. JELLIFFE[18] recommends, nevertheless, that carotene-containing foods "should be given to lactating mothers in regions where vitamin A deficiency diseases are prevalent."

There is general agreement[46, 54, 55] that large doses of vitamin A will elicit increased vitamin A secretion in breastmilk. In a carefully controlled experiment on one woman[46], it was clearly shown that profound effects could be wrought on the vitamin A content of milk by administering a daily dose of as little as 2500 I.U.

A meal of liver can double the vitamin A concentration of milk, a day later. In those parts of the tropics where red palmoil is in daily use as an article of food, it is unlikely that the vitamin A store of a lactating mother would be depleted and hence her milk is not likely to be deficient in the vitamin. This does not imply that large Vitamin A stores could be built up in the mother by dietary means during pregnancy to be used later to significantly raise the vitamin A output in milk[46]. The carotenoid contribution to the vitamin A activity of breast milk is not considered by HENRY and his co-workers as amounting to more than about 2%.

In order to maintain satisfactory haemoglobin synthesis in a baby of six months, STEARNS & STRINGER estimate that a daily retention of at least 0.7 mg of iron is desirable[56]. This is approximately seven times the average daily retention of the metal from breast milk. Yet, nutritional anaemia sets in slowly in breast-fed infants. Cows' milk provides practically no retention of iron by an infant. The folic acid present in milk is possibly of use in enhancing the utilization of the iron[41]. However, the extent of nutritional anaemia in the tropics is emphasized by STRANSKY & DAVIS-LAWAS[57] who found in all of 120 poorly nourished Filipino children, aged between 6 months and 2 years, 10 g or less haemoglobin per 100 ml blood, by WELBOURN[58] whose investigation showed the Buganda children in East Africa to have a much lower blood haemoglobin level than English infants of comparable age, with a deficiency of about 1 g/100 ml on the average; and also by JELLIFFE, WILLIAMS & JELLIFFE[20] who found an average haemoglobin concentration of 9.8 g/100 ml of blood of poorly nourished Jamaican village infants. But for the widespread practice of cooking green vegetables in Asia and Africa before consumption, enough vitamin C could probably be obtained from the diet to provide the supply which a breast-fed child needs.

JEANS[41] considers 30 mg ascorbic acid daily as ample supply for a six-months old baby. Slightly less is required earlier. The amount of vitamin C which a breast-fed baby obtains from its mother's milk depends upon the maternal intake of ascorbic acid[46, 59, 60]. On saturation of the mother's tissues with the vitamin, the concentration of ascorbic acid has been shown to be within the limits of 5 mg and 11 mg/per 100 ml milk[61, 62, 63]. When the upper limit is reached, any additional intake of vitamin C is excreted in the urine[46]. Similarly, when the mother's food is deficient in vitamin C, the urinary excretion falls more quickly than the milk value. Hence the milk ascorbic acid content is a good pointer to the mother's state of nutrition.

LESNE & BRISKAS[64] quote a range of 0.6—0.95 mg/ml in the 2nd month and 0.3—0.7 mg/ml in the 9th month. Human milk ascorbic acid is very labile and has been shown to lose as much as 80% of its original potency after standing six hours at 37° C in the dark. Only 15% loss occurs in cow's milk[46]. It is possible that the increased lability of human milk is due to its higher content. HARRIS & RAY[119] obtained a value of 5.6 mg/100 ml milk of 13 English women. The work of HENRY et al.[46], with 14 samples of English milk, yielded an average lying between 2.5 and 4 mg/100 ml. Very little quantitative data on the seasonal variation of Vitamin C in the milk of tropical peoples have been gathered.

SQUIRES[65] analysed the milk of 84 malnourished Tswana women in Buchuanaland and obtained a mean value of 1.7 mg ascorbic acid/100 ml in the dry season and 2.7 mg/100 ml in the rainy season when fresh vegetables and fruit were freely consumed by the mothers. Similarly, STRANSKY, DAVIS-LAWAS & VICENTE[66] found low concentrations of vitamin C in the milk of lactating mothers during the first year. In 30% of the cases studied the amounts were "extremely low."

It is surprising, therefore, that clinical scurvy in infants is extremely rare[18, 66, 67]. Care has to be exercised in assessing the value of the results of ascorbic acid estimation in human milk because of the high lability of the vitamin[46]. And it may well be that these reports of low ascorbic acid content of the breast milk of tropical women are due to under-estimation consequent upon the spontaneous oxidation of the vitamin.

Artificial Feeding

In general, artificial feeding becomes necessary when (a) the mother's lactation fails or is inadequate, (b) the infant is losing weight or not gaining, (c) the child fails to thrive as a result of some dietary deficiency in its mother's milk and (d) it is desired to train the infant towards weaning and being omnivorous.

Cow's milk may be fed to a baby as a complement to its breast milk feeds. Supplementary feeding of vitamin and mineral rich foods is frequently recommended. In both cases, it is essential that a condition of near asepsis prevails around the handling of the infant's food, especially in the tropics where intestinal infection is the chief killing disease of childhood[18].

The energy requirement of individual children varies, and so only approximate caloric requirements can be made. It is usual to assume that the caloric equivalents of the main items of food are as follows: – carbohydrate, 4.1 calories per gram; protein, 4 cal/g; fat, 9.3 cal/g. The caloric requirements of babies in the tropics is slightly less than that of European and North American babies. The latter is given to be between 100—110 calories per kilogram body weight per day[68], for the first five months of a child's life. In hot countries an estimate of 90—100 cal/kg is, perhaps, the optimum. When food is ample and the baby is feeding well, a fourfold increase in its caloric intake may be experienced in the first four days of life, so that the value of the food swallowed rises from about 22 cal/kg to 88 cal/kg. A further rise to above 100 cal/kg body weight/ day may occur. However, after the fifth month the caloric requirement is usually stabilized at a figure lower than the 4-month peak.

In determining the amount of supplementary or/and complementary feeds that should be given to a child, the metabolic type of the child has to be borne in mind. Small, restless children need more calories than fat, placid ones. EVANS & MACKEITH[69] have criticized the so-called normal weight charts on the grounds that both the birth weight and the body weight at different periods in childhood show individual characteristics. It is possible that a racial factor is also involved. It seems more intelligible to consider the ranges of body weight at the various times as the true yard-stick. For English children, LIGHTWOOD[68] notes 5—12½ lbs at birth, 10—16 lbs at three months, 14—20 lbs at six months, 16—22 lbs at nine months, and 18—24 lbs at twelve months. These data may be compared with those given by KROGMAN[118] p. 134, for Scottish children.

It is of doubtful value to attempt, by giving prelacteal feeds, to replenish the loss of weight which inevitably occurs in the first two days of life when the infant is drawing its energy mainly from its glycogen stores[70-76]. The present-day view is that only sterile water should be given at this time unless the fall in weight is greater than 10% of the birth weights[2, 68] KUGELMAN, BERGGREN & CUMMINGS[77] maintain that the average weight loss of babies in this initial period can be reduced from 7% to 1.7% by feeding, at 2 hourly intervals, a mixture of dextrose gelatine and sodium chloride. ILLINGWORTH is concerned with the effect which prelacteal feeds have on the establishment of lactation[2] and the onset of oedema in the infant[75]. JELLIFFE[18] strongly recommends that no complementary feeds should be given to infants in tropical countries during the first six months of life, owing to the grave risk of the baby being infected in the process of feeding. However, where the milk supply is inadequate to support optimum growth, or extra vitamins are required, it would be foolhardy to confine infant feeding to the sucking of the breast.

While it is not now considered absolutely necessary that an artificial feed should correspond exactly with the composition of breast milk[68] it is, however, desirable that the diet be fed to an infant in fluid or semi-fluid form. HOLT[124] is of the opinion that it is probably useless to attempt to better human milk formulation by providing additional nutrients in supplementary diets which are made of cow's milk. MACKAY[78] has described standard mixtures which are suitable for feeding artificially to babies. It is generally considered best to make the food mixtures roughly equivalent to 70 cal per 100 g, which is the caloric value (approximately) of both cow's milk (except Jersey and Guernsey) and human milk[63]. Some infants with digestive troubles do better on a more concentrated feed, while others prefer them dilute.

Sugar can replace milk in the preparation of feeds, if it is remembered that six grams of sugar dissolved in 28.5 g water have the same caloric value as milk i.e., 70 cal/100 g. Similarly, dried milk may be replaced or replenished by sugar. Tables VIII and IX illustrate LIGHTWOOD'S[68] recommendation for normal English babies.

In many parts of the Tropics cow's milk is in very short supply and artificial feeding becomes a much more complicated matter. DEAN[86] has reviewed in detail various experiments [81, 85] in the use of plant products for infant feeding.

Table VIII

Feeding Schedule for normal infants fed artificially on mixtures containing Cow's milk

Age in Months	1	2	3	4	5	6
Approximate weight in pounds	8.5 (3868)	10.5 (4776)	12 (5460)	13.5 (6143)	15 (6835)	16 (7280)
Number of feeds in 24 hours	5 or 6	5	5	5	5	5
Amounts of Cow's Milk in ounces	2.5 or 3 (71 or 86)	4 (114)	5 (142)	5.5 (157)	6.5 (185)	7 (199)
Amounts of sugar in level tea spoons	1.5	1.5	1.5	1.5	1.5	1.5
Boiled water in ounces	1 (28.5)	1 (28.5)	1 (28.5)	1 (28.5)	1 (28.5)	1 (28.5)

Derived from a paper by LIGHTWOOD[68] by the courtesy of the publishers, Butterworths Scientific Publications, London.
The figures in brackets are the corresponding values in grams.

Table IX

Feeding Schedule for normal infants fed on mixtures
containing Full-Cream National Dried Milk.

Age in Months	1	2	3	4	5	6
Approximate weight in pounds	8.5 (3868)	10.5 (4776)	12 (5460)	13.5 (6143)	15 (6835)	16 (7180)
Number of feeds in 24 hours	5 or 6	5	5	5	5	5
No. of level measures of milkpowder	2.5 or 3	4	5	5.5	6.5	7
No. of level measures of sugar per feed	1.5	1.5	1.5	1.5	1.5	1.5
Ounces of boiled water	3.5 or 4 (100-114)	5 (142)	6 (171)	6.5 (185)	7.5 (214)	8 (228)

After LIGHTWOOD[68]. Reproduced by the courtesy of the publishers, Butterworths
Scientific Publications, London.

The figures in brackets are the corresponding weights in grams. The "measures"
referred to in this table is supplied with the National Dried Milk.

DEAN's own experiments, which involved the use of four types
of diets compounded around soya bean powder, are illustrated in
Table X. Variants of the main types of mixtures were used:

B_1 contained 87.5 parts B + 12.5 parts skimmed milk powder
B_2 contained 75   parts B + 25   parts skimmed milk powder
C_1 contained 90   parts C + 10   parts skimmed milk powder
C_2 contained 10   parts dried milk solids and was obtained
by addition of fresh skimmed milk to mix C before drying.

D_1 and D_2 were the same mixture as D, except that one was made
in the laboratory and the other in the factory.

Both the method and results of DEAN's feeding experiments are
interesting and will now be described. The children used in the
experiments lived in two Roman Catholic children's homes in
Germany and were divided into four groups: 6—12 months old,
1—2 years, 2 to 5 years and 7 to 11 years. The original diet of the
children was a brei based on wheat semolina. The milk used in the
preparation of the brei was usually skimmed and vitamin supplements were rarely given to the children. In DEAN's experiments,
the soya mixtures were reconstituted with water and used in the
preparation of brei for the children. Oily solutions of vitamins A and
D were usually added to the brei at mid-day together with one or
more of the vegetables carrots, cabbage, potato, apple pulp, tomato
pulp. Vitamin C was given either as orange juice or as crushed

14*   (211)

Details of Soya Mixtures used by DEAN[86] in Child feeding experiments. Reproduced from
and the Controller of Her Brittanic

Mixture	Trypsin in-hibitor	Total N (g 100g)	Protein (g 100g)
A. 9.6 kg national flour (85% ext.) heated for 15 mins. in 50 litres water almost to boiling: stir to gelatinize starch and cool to 70 °C by adding cold water. Keep at 63 °C till all starch hydrolysed. Then slowly add 9.6 kg of Soya flour containing 20% fat plus 0.64 kg NaCl and 2.7 kg Calcium lactate and mix. Dry in vacuo between 4.5 °C and 50 °C.	Not Des-troyed	1.61	9.3
B. 89 kg barley flour and 11 kg wheat flour (70% ext.) mixed with 450 l water and heated at 50 °C for 45 mins. Then at 70 °C to hydrolyze all starch. 100 kg soya beans steamed for 50 mins. dried and ground sieved. Suspend powder in H_2O. This added to 80 kg malt solids in suspension and the whole concentrated to 75% solids in vacuo and spray dried.	B B_1 B_2 Partly destroyed	2.95 3.28 3.6	17.6 19.2 21.4
C. Malt extract made as for mix B. filtering & evap. under reduced pressure to get 78% solids. Five parts soya powder prepared as for B are suspended in 3 vols. H_2O and steam passed through for 120 mins. Now mix the malt ext. and soya suspension with some calcium carbonate. Dil. with H_2O and spray dried.	All Des-troyed. C_1 C_2	2.93 3.22 3.20	17.5 18.7 18.6
D. Prepared as for Mix C but cow's milk added before drying to give 10% milk solids in the dry product. The malt ext. made from equal parts of barley flour & maize flour. Sucrose added to the suspension before drying so as to obtain 7.5% additional sugar in the dry product.	Des-troyed Des-troyed	3.0 2.92	17.4 17.0

ascorbic acid. The weight of food and its constitution was determined before and after every meal and the amount consumed noted.
The first experiment with the youngest children lasted 12 weeks.
The first six weeks were a transition period in which the test diet
was gradually introduced. In the next six weeks only the experimental diet was used. In all 25 infants took part in the test. They
were divided into approximately, six in a group for each test diet.
With the children aged 1—2 years the experimental periods were

X

Medical Research Council Special Report Series No. 279, Table 15, by courtesy of the author Majesty's Stationary Office.

Fat (g/ 100g)	Cal- cium (mg/ 100g)	Phos- phorus (mg/ 100g)	Magne- sium (mg/ 100g)	Iron (mg/ 100g)	Vita- min B (mg/ 100g)	Calor- ies /100 g	
2.6	631	345	86	1.8	0.46	392	
7.5	211	457	157	9.1	0.52	423	B
6.6	342	532	162	8.1	—	417	B_1
5.7	747	607	146	6.9	—	415	B_2
7.3	556	489	—	—	—	421	C
—	—	—	—	—	—	417	C_1
6.5	778	506	—	—	—	417	C_2
—	—	—	—	—	—	416	D_1
6.3	750	485	—	—	—	416	D_2

16 and 24 weeks. There were 50 experimental subjects in this group and they were divided into rather larger sub-groups than were the youngest children. Three trials were conducted with a total of 50 subjects in the age group 2—5 years. In the oldest group, the trials began with a total of 125 children, but many did not complete the test. A group of 25 newborn babies were also studied. In each trial, a control group receiving a supplement of only milk was kept.

Measurements of heights, weights and blood haemoglobin were

52

made before and during the entire trial period. Assessment of the nutritional value of each experimental diet was made by two criteria:

(1) actual gain in weight during the trial period, minus
   Standard gain in weight during the trial period
(2) (% of the accepted standard of body wt. at start of trial) minus
   (% of the accepted standard of body wt. at end of trial)

The standards accepted were obtained after consideration of data provided by BOYNTON[87], MEREDITH[88] and O'BRIEN, GIRSCHICK & HUNT[89].

In general, DEAN's results showed that plant products can be made into adequate substitutes for milk in human nutrition. It was shown that half the milk diet of children less than one year old can be replaced by the soya mixtures fed. At the age of 1—2 years, it seemed as if all the dietary milk can be replaced in this way by two of the mixtures. All the mixtures gave satisfactory results with the children over two years of age, but for optimum growth a small amount of milk was required in addition to the soya mixtures. In almost all cases the children put on weight during the course of the trials.

In view of the dangers of infection which may be attendant upon the giving of supplementary foods to young infants in the Tropics[18], BASSIR[117] has reported observations on 16 Nigerian mothers in the first six months of lactation. The subjects were selected at random from those attending an infant welfare clinic in Lagos and then divided into two groups according to their stage of lactation and socio-economic status. One group received a supplement of 30 grams soya flour (equivalent to 14.5 g protein) in addition to their daily customary diets. An increase in the output of breast milk was obtained after 4 weeks and became much higher after 8 weeks. After one month the sample variance was 66.6 and the value for "t" was 2.65 but the concentration of the proximate principles, including phosphatase, were not affected. These results are in agreement with the view of ANTONOV[9] that only the quantity and duration of lactation are affected by dietary restriction, and they do point a way to improving infant feeding in cases were the nutrition of the mother is inadequate.

The giving of the oils of cod, halibut and shark is recommended as a routine in supplementing infant feeding, because of the importance of the vitamins A and D they contain. The amounts required and the rationale for their administration has been discussed above. Orange juice may be given to infants from the earliest days of its life, as a source of vitamin C. Many other plant products, e.g. rose hip syrup and black currant syrup are also available in commercial preparations.

Another kind of supplementary food which has long been in

common use in the tropics, for supplementary feeding of infants, is goat's milk. Its main advantage over cow's milk is the fact that diarrhoea seldom results from its use. Goat's milk appears to contain a factor (as yet unidentified) which is anti-infective.

With all supplementary feeding, over-nutrition is a danger to be guarded against. It is unlikely that this would become a major problem in the Tropics for a long time to come, since the general standard of living is low and most of the countries are very poor. The solution is to calculate the food intake required by supplementary dieting, on the basis of the mean derived from the actual weight of the infant and that expected at its stage of life. In this way an under-weight baby will get a little more than his requirement and so be in a better position to make up the deficiency; and an over-weight child will get just that little less than its optimum requirement and hence gradually return to normal weight.

Fluid Intake, Water Loss and Antimetabolites

All other things being equal, young infants have a higher percentage of body water than older ones. 70% of the body of new-born infants is composed of water. Premature babies have a higher percentage still. From the age of six months till the onset of adolescense the total body water remains constant around 60%[90, 91, 115]. About 4% of the body wt. is plasma, with erythrocytes.

In tropical conditions body water is of immense importance in the maintenance of electrolyte balance, in the excretion of waste material, in controlling body temperature and cerebro-spinal fluid pressure, and in maintaining an even blood volume.

Only one-third of the body water is normally extracellular and about 75% of this is in the interstitial fluid. The distribution of fluids in the body is closely related to the concentration of dissolved proteins and electrolytes. Water is generally dispersed in a discontinuous phase in cells and cell colloids, thus determining the firmness of the gel state of the colloid.

All children are susceptible to dehydration because they are apt to lose much water through perspiration, vomiting, faeces and urine. And the intake of water to correct these losses may reach very high proportions. EBBS[91] puts the volume of water an infant takes in its food and drinks as amounting to six times the volume required by an adult, per unit body weight. GOLDBLOOM[92] reckons it as 187.5 ml H_2O/kg body weight.

In the West Indies, West Africa, and other parts of the Tropics dehydration of infants is prevented through the administration of frequent drinks of vegetable infusions and water.

When body water is lost by a baby, it is generally accompanied

54

by sodium and chloride. If only the latter is lost, e.g. in excessive vomiting, alkalosis may result. This would tend to push the level of serum calcium down and thus to increase the risk of tetany. Increased viscosity of the blood plasma may lead to renal failure. When excessive amounts of bases have been lost, acidosis may ensue.

The correction of electrolyte imbalance and kidney dysfunction in dehydrated infants requires great care and a good deal of biochemical knowledge. The aim is always to restore the balance of anions and cations at the correct physiological level. The homoeostatic quality of the kidney is such that no special measure is ordinarily required to bring the kidney function to normal if the electrolyte balance has been restored.

It is probably best to administer a 10% glucose or molar lactate solution in 0.88% saline through a nasal catheter which reaches down the oesophagus. Parenteral drips of about 20 drops a minute through intravenous medium is useful in urgent cases, but may be dangerous if frequent biochemical tests are not performed on the infants blood during the treatment. Rough clinical assessments, e.g. feeling the turgidity of the anterior fontenelle are only of limited value.

Calcium may be lost from the alimentary canal through excessive vomiting. When this results in the transfer of calcium ions from the skeletal tissues to maintain the normal concentration of serum calcium, there is a risk that, when the electrolyte balance of sodium and chloride has been corrected, calcium ions will be deposited in the bones faster than can be replaced through dietary means. However, this may soon be corrected as more and more milk feeds are assimilated.

There is strong indication[93, 94, 98, 99] that, when babies are given vegetable infusions to correct or prevent dehydration, not only could the need for water be exaggerated, but liver function may be impaired in a manner which is reflective of diminished antidiuretic action of the pituitary hormones. HILL and his co-workers[95] found hepatosis, ascites, tender liver, and cirrhosis in Jamaican infants who were improperly fed on low protein and refined cereals, and they thought that the "bush tea" given to the babies was an aggravating factor. COOK, DUFFY & REGINA SCHOENTAL[96] had produced similar examples of liver damage in rats fed with *Senecio* alkaloids. *Senecio* plants are of wide distribution in the Tropics and are reported to be favourite remedies of the African Negroes, who use them for a variety of disorders, including charming away bad dreams of babies. SCHOENTAL[97] is therefore of the opinion that it would be profitable to find out whether the liver damage described by HILL and his colleagues gives rise to liver carcinoma in adult Africans, as is known to happen in experimental animals.

(216)

In some cases, when lactation fails or the infant is unable to remove sufficient milk, as much as half the body weight of a four-months old baby may be lost in a week through sweat, urine and vomiting; and this loss may be made up in the space of a fortnight by feeding dilute saline and milk by mouth[100].

GILLMAN & GILLMAN[101] have reviewed, in detail, factors relating to the retention of water in nutritional oedema. There is no doubt that hypoproteinemia is the most consistent feature of the condition. According to EMDIN[102] protein deficiency due to the body's inability to absorb the nutrient from the gut is probably the precipitating cause. In Kwashiorkor, oedema is usually accompanied by gastro-enteritis, persistant vomiting, low plasma albumin and fatty liver diseases[26, 102]. SCHNIEDEN et al.[121] have confirmed by means of deuterium an increase in total body water in Kwashiorkor children in Nigeria. But there is as yet no agreement as to whether this increase is due to extra-cellular or intra-cellular hypotonicity, or to both[122, 123].

Both in hunger oedema and Kwashiorkor, there occurs a metabolic derangement which involves potassium, proteins, inorganic phosphate, calcium and sodium chloride of blood[120]. An indication of this complex inter-relationship is given by PASSARO[103, 104] in his experiments with nine oedematous children (11—24 months old), who had been kept on a high carbohydrate-low protein diet for periods up to five months. Values of over 41% body weight were obtained for the extra-cellular fluid content of the children, and these fell by a third, on average, as the electrolyte imbalance was rectified in five of the children. During treatment of five of the children a marked change of the electrolyte composition of the blood towards the normal was noted. The following average values in m. eq/l were obtained for plasma:

	Sodium	Potass ium	Cal- cium	Chlo- ride	Inor- ganic P	Pro- tein	Total Base
Early in treatment	124.1	3.7	4.4	91.8	1.8	10.7	135.3
3-5 weeks later	136.3	4.5	4.9	101.5	2.8	16.6	148.8

Digestion and Absorption of Nutrients

Malnutrition in tropical infants is often associated by gross upset in the gastro-intestinal organs so that the functions of these organs fall far below normal. JELLIFFE[18] has summarized the main features of this kind of disorder in the disease Kwashiorkor:

"Anorexia and vomiting are common, and loose, rather bulky stools containing undigested food are usual. A moderate steatorrhoea may be present. The loose stools would appear to have the following etiology: (a) diminished exocrine pancreatic secretion as a result of acinar change in the pancreas; (b) intestinal malabsorption due to decreased secretion of the enzymes of the succus entericus and possibly to actual changes in the intestinal mucosa; (c) associated gastro-enteric infection; (d) continued ingestion of unsuitable indigestible foods, associated with intolerance of fat, including milk fat, and of sugars, including lactose."

By means of objective biochemical analysis, it is possible to test the efficiency of gastric and pancreatic function, and hence the veracity of JELLIFFE's statement.

WATERLOW[105], using a mixture of sugar, water and brandy as stimulus, did not observe, at half hour intervals, any marked deviation from the normal amounts of free and total gastric acidity in the seven children with fatty liver disease whose juices were examined in the West Indies. But there was a tendency towards decreased acidity after recovery or improvement. On the other hand, JAYASEKERA, DE MEL & COLLUMBINE[106], who gave test meals to their Kwashiorkor infants, reported free acid concentration at 60 minutes

after the meal, ranging from 0—46 ml $\frac{N}{10}$ HCl.

Peptic activity was also measured by JAYASEKERA et al.[106] in all of their eight subjects, but the results did not reveal a corresponding impairment of enzyme function.

In so far as the exocrine function of the pancreas is indicated by the amount of lipids present in the faeces of infants, the work of JAYASEKERA and his associates[106], VAN DER SAR[107], THOMPSON[108], [110], [111] and DEAN[109] all suggest impairment of lipolytic activity in infants suffering from protein malnutrition. In 17 out of 19 cases studied by VAN DER SAR[107], over 30% of the dry weight of faeces consisted of lipids. The mean percentage of total fats in the faeces of Kwashiorkor infants studied by JAYASEKERA et al.[106] was three times the normal amount in the cases which survived their illness and four times the normal, when the disease proved fatal. 10.0%, the mean faecal fat content of children who were well-fed, was regarded as normal. TREMOLIERES[126] has shown by the method of differential centrifugation that the amount of cellulose fibres in the faeces depends not only on the dietary intake of this carbohydrate, but also on the amount of fat to be excreted. The fact that children suffering from Kwashiorkor pass out rough bulky stools can thus be explained.

THOMPSON & TROWELL[111] showed that the concentration of lipase rose to approximately the same values as those of control infants, when infants suffering from protein-deficiency disease were successfully treated with high protein diets. Comparing the effects of vegetable and animal proteins fed at different levels to Kwashiorkor infants, MARGARET THOMPSON[110] found that in resting duodenal contents the lipase activity reached 20—50 units/ml juice while four children were on a milk diet containing less than 20 g protein per day. At comparable levels of maize protein, the highest activity was 10 units per 1 ml juice. The result of feeding soya and milk diets, at a level of 4 g protein/kg body weight, to each of two groups of five Kwashiorkor infants was to raise the mean lipase activity from 9 and 11 units/ml juice before the diet to 63 and 79 units/ml juice, respectively, after the diet. The unit of lipase in this experiment is numerically equal to the number of ml $\frac{N}{20}$ alcoholic NaOH required to titrate the fatty acid which is liberated from 1.0 ml olive oil by the enzyme contained in 0.1 ml pancreatic juice, in the presence of 0.1 ml 25% bile salt solution and 5 ml phosphate buffer of pH 8.0.

Determinations of trypsin activity and amylase activity were concurrently made on the same infants with results of roughly the same order as those of lipase.

Table XI

Trypsin Activity of resting Duodenal Juice of East African children

Controls without Kwashiorkor Symptoms		Kwashiorkor Infants		
Age (months)	Trypsin Activity (units/ml)	Age in months	Trypsin Activity (units per ml)	
			Before Treatment	After Treatment
		24	3.5	19.5
24	24	18	1.5	22.5
12	17	21	3.0	9.5
15	45	36	3.5	22.5
36	15	30	7.0	14.0
42	13	51	3.5	8.5
18	6.5	18	2.5	20.0
		18	5.0	22.0
		18	5.0	22.5
		18	2.5	24.0
Mean =	20.1	Mean =	3.7	18.5

Derived from a paper by THOMPSON[110] and reproduced by courtesy of the publishers.

58

Amylolytic action was assessed from the speed of hydrolysis of starch in the manner of AGREN & LAGERLOF[112].

The methods of CHARNEY & TOMARELLI[113] and TOMARELLI et al.[114], based on the enzymatic digestion of Sulphanilamide — azocasein or azoalbumin, were employed in THOMPSON's[110] estimation of the tryptic activity in resting duodenal contents of East African children. Her results are summarized in Table XI, and well illustrate the low trypsin concentration in the duodenal juice of infants suffering from protein deficiency disease. It can be seen that in every case, the enzyme activity is at least doubled as a result of treatment with high protein diets.

REFERENCES

1. NAISH, C., 1953. Paediatrics, **1**. Butterworth Sci. Publ., London.
2. ILLINGWORTH, R. S., 1953. The Normal Child, Churchill, London.
3. WALLER, H., 1950. *Lancet*, **i**, *53*.
4. MOLONEY, J. C., 1945. *Psychiatry* **8**, *391*.
5. NEWTON, M. & NEWTON, N. R., 1948. *J. Pediat.* **33**, *698*.
6. BASSIR, O., 1956. *W. Afr. med. J.* **5**, *88*.
7. TOVERUD, G., 1950. *Millbank Mem. Fund Quart.* **28**, *7*.
8. ABELS, J., 1949. *Brit. med. J.* **i**, *154*.
9. ANTONOV, A. N., 1947. *J. Pediat.* **30**, *250*.
10. SMITH, C. A., 1947. *J. Pediat.* **30**, *229*.
11. WALKER, A. R., ARVIDSSON, ULLA & DRAPER, W. L., 1952. *Lancet* **I**, *317*.
12. KROPMAN, M., 1946. M. Sc. thesis, University of S. Africa.
13. BASSIR, O., 1956. *J. trop. Med. Hyg.* **59**, *139*.
14. TROWELL, H. C., 1948. *E. Afr. med. J.* **25**, *236*.
15. WIESCHHOFF, H. A., 1940. *Bull. Hist. Med.* **8**, *1403*.
16. SPENCE, J. C., 1938. *Brit. med. J.* **ii**, *729*.
17. CANET, J., 1952. *Rev. Colon. Méd. Chir.* **24**, *2*.
18. JELLIFFE, D. B. 1955. Infant Nutrition in the Subtropics and Tropics. W. H. O. Geneva.
19. VERGARA, A., SANTOS, W. & WATERLOW, J. C., Personal Communication.
20. JELLIFFE, D. B., WILLIAMS, L. L. & JELLIFFE, E. F., 1954. *J. trop. Med. Hyg.* **57**, *27*.
21. MEAD, MARGARET, 1935. Sex and Temperament. Routledge, London.
22. MEAD, MARGARET, 1949. Male and Female. Gollancz, London.
23. WOODRUFF, A. W., 1956. Paediatrics for the Practitioner. Butterworths Sci. Publ., London.
24. KON, S. K. & MAWSON, E. H., 1950. Human Milk, M. R. C. Special Report Series No. **269**.
25. DEAN, R. F., 1951. Studies in Undernutrition M. R. C. Spec. Rep. Series no. **275**.
26. TROWELL, H. C., DAVIES, J. N. & DEAN, R. F., 1954. Kwashiorkor. Arnold, London.
27. HYTTEN, F. E., 1954. *Brit. med. J.* **i**, *175* and *249*.
28. AUFFRET, C. H. & TANGUY, F., 1949. *Bull. Med. A. O. F.* **6**, *99*.
29. BASSIR, O., 1954. *Proc. Nutrit. Soc.* **13**, abs. xv.
30. BIGWOOD, E. J., 1954. in: Malnutrition in African Mothers, Infants and Young Children. H. M. S. O., London.

59

31. BLOCK, R. J. & MITCHELL, H. H., 1946. *Nutr. Abs. Rev.* **16**, *249*.
32. MACY, I. G., 1949. *Amer. J. Dis. Child.* **78**, *589*.
33. SRINIVASAN, P. R. & RAMANATHAN, M. K., 1954. *Indian J. med. Res.* **42**, *51*.
34. GOPALAN, C., 1956. *J. trop. Paediat.* **2**, *89*.
34. NICHOLLS, L., 1951. Tropical Nutrition. Bailliere, Tindall and Cox, London.
35. UGA, Y. 1935. *Tohoko J. exp. Med.* **25**, *169*.
36. CAWLEY, R. H., McKEOWN, T. & RECORD, R. G., 1954. *Brit. J. prev. soc. Med.* **8**, *66*.
37. BROCK, J. F. & AUTRET, M., 1952. *Bull. W. H. O.* **5**, *1*.
38. NICOL, B. M., 1952. *Brit. J. Nutrit.* **6**, *34*.
39. DEAN, R. F. & SCHWARTZ, R., 1954. *Courrier*, **4**, *293*.
40. WIDDOWSON, E. M., 1950. *Nature, Lond.* **166**, *626*.
41. JEANS, P. O., 1951. Handbook of Nutrition. H. K. Lewis, London.
42. STEARNS, G., JEANS, P. C., & VANDECAR, V., 1936. *J. Pediat.* **9**, *1*.
43. STEARNS, G., 1939. *Physiol. Rev.* **19**, *415*.
44. POLSKIN, L. J., KRAMER, B. & SOBEL, A. E., 1945. *J. Nutrit.* **30**, *451*.
45. OUTHOUSE, J., MACY, I. G. & BREKKE, V., 1928. *J. biol. Chem.* **78**, *129*.
46. HENRY, K. M., KON, S. & MAWSON, E. H., 1950 in: Human Milk. M. R. C. Spec. Rep. Series No. **269**.
47. DRUMMOND, J. C., GRAY, C. H. & RICHARDSON, N. E., 1939. *Brit. med. J.* **ii**, *757*.
48. JEANS, P. C. & STEARNS, G., 1938. *J. Amer. med. Ass.* **711**, *703*.
49. HESS, A. F. & WEINSTOCK, M., 1927. *J. Amer. med. Ass.* **88**, *24*.
50. DANN, W. J., 1940. *Biochem. J.* **34**, *724*.
51. FRIDERICHSEN, C. & WITH, T. K., 1939. *Ann. Paediat.* **153**, *113*.
52. McCOSH, S. S., MACY, I. G., HUNSCHEN, H., ERIKSON, B. N. & DONELSON, E., 1934. *J. Nutrit.* **7**, *331*.
53. WITH, T. K. & FRIDERICHSEN, C., 1939. *Ugeskr. Laeg.* **101**, *915*.
54. HRUBET, M. C., DEVEL, H. J. & HANLEY, B. J., 1945. *J. Nutrit.* **29**, *245*.
55. NEUWEILER, W., 1935. *Z. Vitaminforsch.* **4**, *259*.
56. STEARNS, G. & STRINGER, D., 1937. *J. Nutrit.* **13**, *127*.
57. STRANSKY, E. & DAVIS-LAWAS, D. F., 1948. *Ann. Paediat.* (Basel), **171**, *139*.
58. WELBOURN, H. F., 1955. *E. Afr. med. J.* **32**, *401*.
59. KASAHARA, M. & KAWASHIMA, K., 1936. *Z. Kinderheilk.* **58**, *191*.
60. SINKKO, E. I., 1937. *Acta paediatr., Stockh.* **21**, *407*.
61. WIDENBAUER, F. & KUHNER, A., 1938. *Z. Vitaminforsch.* **6**, *50*.
62. GEAHTGENS, G. & WERNER, E., 1937. *Arch. Gynaek.* **165**, *63*.
63. BAUMANN, T., 1937. *J. Kinderheilk.* **150**, *13*.
64. LESNE, E. & BRISKAS, S., 1938. *Acta paediatr., Stockh.* **22**, *123*.
65. SQUIRES, B. T., 1952. *Trans. R. Soc. trop. Med. Hyg.* **46**, *95*.
66. STRANSKY, E., DAVIS-LAWAS, D. F. & VICENTE, C., 1950. *J. trop. Med. Hyg.* **53**, *170*.
67. FERNANDO, P. B. & RAJASURIYA, P. K., 1946. *Indian J. Pediat.* **13**, *69*.
68. LIGHTWOOD, R., 1953. Paediatrics, **1**. Butterworths Sci. Publ., London.
69. EVANS, P. R. & MacKEITH, R., 1951. Infant Feeding. Churchill, London.
70. SANFORD, H. N., 1937. *J. Amer. med. Ass.* **117**, *470*.
71. BRUCE, J. W., 1936. *J. Pediat.* **8**, *651*.
72. KALISKI, S., 1941. *Texas State J. Med.* **37**, *282*.
73. PARMOLEE, A. H., 1936. *J. Pediat.* **8**, *646*.
74. RODDA, F. C. & STRESSER, A. V., 1938. *Wisconsin med. J.* **37**, *547*.
75. KROST, G. N. & EPSTEIN, I. M., 1931. *J. Pediat.* **10**, *221*.
76. SCHORER, E. H. & LAFFON, F. L., 1935. *J. Pediat.* **17**, *613*.
77. KUGELMAN, I. N., BERGGREN, R. E. & CUMMINGS, M., 1933. *Amer. J. Dis. Child.* **46**, *280*.
78. MACKAY, H. M., 1941. *Brit. med. J.* **1**, *841, 888*.
79. WALTERS, J. H. & WATERLOW, J. C., Fibrosis of the liver in W. african Children. M. R. C. Spec. Rep. Series No. **285**.

60

80. HOLT, L. E. & McINTOSH, R., 1941. Revision of Holt's Diseases of Infancy and Childhood, 11th edition. New York.
81. PLATT, B. S., 1936. *Chin. med. J.* **50**, *410.*
82. FAN, C. WOO, T. T. & CHU, F., 1940. *Chin. med. J.* **58**, *53.*
83. DESIKACHAR, H. S. & SUBRAHMANYAN, V., 1949. *Indian J. med. Res.* **37**, *77.*
84. KARNANI, B. T., DE S. S., SUBRAHMANYAN, V. & CARTNER, D., 1948. *Indian J. med. Res.* **36**, *355.*
85. GESTEIRA, M. & BAHIA, A., 1932. *Brazil-Med.* **46**, *173.*
86. DEAN, R. F., 1953. Plant Proteins in Child Feeding. M. R. C. Spec. Rep. Series No. **279.**
87. BOYNTON, B., 1936. Univ. Iowa Stud. Child. Welf. No. **325.**
88. MEREDITH, H. V., 1935. Univ. Jawa. Stud. Child. Welf. No. **292.**
89. O'BRIEN, R., GIRSCHICK, M. A. & HUNT, E. P., 1949. Misc. Publ. U.S. Dept. Agric. No. **366**, Washington.
90. FRIIS-HANSEN, B. J, HOLIDAY, M., STAPLETON, T. & W. WALLACE, W. M., 1951. *Pediatrics,* **7**, *321.*
91. EBBS, J. H., 1953. Paediatrics, Vol. **1.** Butterworth, Sci. Publ., London.
92. GOLDBLOOM, ALTON, 1953. Paediatrics, Vol. **1.** Butterworth, Sci. Publ., London.
93. BASSIR, O., 1953. *W. Afr. med. J., N. S.* **2**, *30.*
94. HELLER, H., 1953. Quoted by TROWELL et al. 26.
95. HILL, K. R., RHODES, K., STAFFORD, J. L. & AUB, R., 1953. *Brit. med. J.* **1**, *117.*
96. COOK, J. W., DUFFY, E. & SCHOENTAL, R., 1950. *Brit. J. Cancer,* **4**, *405.*
97. SCHOENTAL, R., 1954. *Brit. med. J.* **1**, *335.*
98. GOPALAN, C., 1950. *Lancet* **1**, *304.*
99. VENKATACHALAM, P. S., SRIKANTIA, S. G. & GOPALAN, C., 1952. *Indian J. Pediat.* **19**, *165.*
100. WATKINS, S., 1955. Personal Communication.
101. GILLMAN, J. & GILLMAN, T., 1951. Perspectives in Human Malnutrition. Grune and Stratton, New York.
102. EMDIN, W., 1953. Paediatrics, Vol. **1.** Butterworth, Sci. Publ. London.
103. PASSARO, G., 1953. *Bol. Soc. ital. Biol. sper.* **29**, *245.*
104. PASSARO, G., 1953. *Bol. Soc. ital. Biol. sper.* **29**, *248.*
105. WATERLOW, J., 1948. Fatty Liver Disease in Infants in the British West Indies. M. R. C. Sp. Rep. Series No. **263.**
106. JAYASEKERA, H. T., DE MEL, B. V. & COLLUMBINE, H., 1951. *Ceylon J. med. Sci.* **8**, *1.*
107. VAN DER SAR, A., 1951. *Docum. Neerl. Indones. Morb. Trop.* **3**, *25.*
108. THOMPSON, M. D., 1952 & 1953. Personal Communication. Kwashiorkor, 1956. Arnold, London.
109. DEAN, R. F., 1954. In: Malnutrition in African Mothers, Infants and Young Children. H. M. S. O., London.
110. THOMPSON, M. D., 1954. In: Malnutrition in African Mothers, Infants and Young Children. H. M. S. O., London.
111. THOMPSON, M. D. & TROWELL, H. C., 1952. *Lancet* **i**, *1031.*
112. AGREN, G. & LAGERLOF, H., 1936. *Acta med. scand.* **90**, *1.*
113. CHARNEY, J. & TOMARELLI, R. N., 1947. *J. biol. Chem.* **171**, *501.*
114. TOMARELLI, R. M., CHARNEY, J. & HARDING, M. L., 1949. *J. Lab. clin. Med.* **34**, *428.*
115. ZWEMER, R. L., 1953. In: Bio Chemistry and Physiology of Nutrition, Vol. 1. Academic Press, New York.
116. ANDERSON, M. & WALKER, A. R. P., 1955. *Brit. J. Nutrit.* **9**, *197.*
117. BASSIR, O., 1959. *Trans. R. Soc. trop. Med. Hyg.* **53**, *256.*
118. KROGMAN, W. M., 1941. Growth of Man. *Tab. Biol.* **20.** Dr. W. Junk Publishers, The Hague.
119. HARRIS, L. J. & RAY, S. N. 1935. *Lancet,* **i**, *454.*
120. SMITH, R. & WATERLOW, J. C., 1960. *Lancet,* **i**, *147.*

121. SCHNIEDEN, H., HENDRICKSE, R. G. & HAIGH, C. P., 1958. *Trans. R. Soc. trop. Med. Hyg.* **52**, *169.*
122. HANSEN, J. D., 1956. *S. Afr. J. Lab. clin. Med.* **2**, *206.*
123. METCALF, J. et al., 1957. *Pediatrics* **20**, *317.*
124. HOLT, EMMETT, L. JR., 1961. Paper in: Recent Advances in Human Nutrition. London. Churchill.
125. BASSIR, O. & CVETKOVIC, J., 1959. *J. trop. Med. Hyg.* **62**, *289.*
126. TREMOLIERES, J., 1961. Paper in: Recent Advances in Human Nutrition. London. Churchill.
127. MORRISON, S. D., 1952. Human Milk. Tech. Comm. No. **18**, Commonwealth Agric. Bureau, Aberdeen.

CHAPTER IV

THE DEPRIVED CHILD (WEANING)

The sense in which the phrase "deprived child" is usually employed is psychological or psychopathological. For the purpose of this study the words are used to indicate the sudden withdrawal of nutritional factors which is, perhaps, the main feature of the weaning of infants in the tropics[1].

That the disease, Kwashiorkor, is caused by this nutritional deprivation at weaning is underlined by the observation of TROWELL, DAVIES & DEAN[2] that nearly half of the total incidence is of children in their second year of life.

Vitamins

While the relationship of protein malnutrition to the incidence and course of the disease, Kwashiorkor, has been the subject of many studies, the biochemical role of vitamins has not been intensively investigated.

Indirect evidence based upon the clinical diagnosis of Xerophthalmia in some Central American children showed other symptoms of Kwashiorkor as vitamin A deficiency with cachexia hidrica and the so called "war oedema"[3, 4, 5].

Attempts to obtain more direct correlation between a dietary lack of vitamin A or its precursors and Kwashiorkor have merely served to illustrate the complicated dysfunction which accompanies the outward signs of the disease[24]. OOMEN[6] who worked with Indonesian children observed eye lesions of the vitamin A deficiency type in children whose dietary intake of β-carotene was assessed as adequate. There was no obvious correlation between the severity of the under-lying Kwashiorkor disease and the ocular Xerosis. When considered in terms of the average normal values for European children, Uganda children, who are severely ill with Kwashiorkor, have been shown by TROWELL, MOORE & SHARMAN[7] to have between 10% and 50% of the expected plasma concentration of carotenoids ([1, 2]). Experiments with 12 Uganda children who had Kwashiorkor averaged 23 mg carotenoid/100 ml plasma while the average values for 4 well-nourished children was 107 mg/100 ml. Four samples of plasma analysed earlier by MOORE & SHARMAN[8] contained only 6, 10, 11 and 25 I.U. Vitamin A/100 ml i.e., 2%, 3.3% 3.7% and 8.3% of the average concentration of plasma carotenoids of well-nourished American children of comparable age range.

It is interesting to note, however, that decreased values of plasma carotenoids have been reported for children with rheumatic fever at various stages of the disease[9]. On the other hand, DITLEFSEN & STOA[10] found the blood level of vitamin A to be increased in persons with chronic liver disease and in alcoholics.

Dark adaptation experiments do not appear to have been conducted in children afflicted with Kwashiorkor although such tests would obviously throw considerable light upon the early phases of the ocular involvement.

TROWELL, MOORE & SHARMAN[7] are among the few who have attempted biochemical studies on possible effects of tocopherol deficiency in Kwashiorkor. They obtained an average of 0.67 mg/100 ml plasma for twelve Kwashiorkor children, as compared with an average of 1.10 mg/100 ml for four normal African children. TROWELL, DAVIES & DEAN[2] have cautiously linked this deficiency with pathological changes in some endocrine glands and in cardiac and voluntary muscles. But the rate of growth and maturity of the gonads are not obviously affected. As there are probably more sensitive criteria, only the most tentative conclusions are as yet possible.

There is a copious literature on animal experiments in which vitamin E deficiency plays a major role, and there is no need to attempt a detailed review of these here. DESSAU, LIPCHUCK & KLEIN[11] produced in mice, by the absence of vitamin E and K, heart muscle lesions similar to those observed by TROWELL et al.[2] in Kwashiorkor, whereas in guinea pig adrenal cortex, PIRTKIEN, STEEGE & ARZT[12] obtained significant increase in vitamin C content after 10 days administration of 0.15 g vitamin E. Successful results were claimed by BUTLER & MCKNIGHT[13] who treated spasmodic primary dysmenorrhoea, among a group of 100 students, with vitamin E. Oral doses of tocopherolesters have been shown to be effective in curing cystic fibrosis of the pancreas in children[25].

Recently more light has been shed on the mode of action of vitamin E by the aid of in vivo studies. With the brown pigmentation of fat observed in vitamin E deficiency as the back-ground of this experiment, TAPPEL[14] demonstrated only a small reduction in oxygen uptake from the addition of tocopherol to an agitated emulsion of linoleic acid and albumin. The brown pigment which resulted from the copolymerization of oxidized fat with protein in such experiments contained 1% to 10% total nitrogen and an uptake of four moles oxygen per mole of linoleic acid. The implication of this work is fortified by the fact that inhibition of autoxidation of unsaturated fatty acids, as well as co-oxidation of carotene and vitamin E, was observed by the same worker[15] using haemoglobin, cytochrome C and haemin as catalysts.

On the immunological side, LAMBERT, SMITH & RICHLEY[16] found that previous injections of peanut-oil and tocopherol in peanut-oil were equally effective in protecting mice against subsequent infection from administered *Salmonella typhi*. It would be interesting to find out whether the vitamin can protect young children whose resistance to bacterial infection is lowered by malnutrition, or if vitamin E deficiency is one of the precipitating causes of the Kwashiorkor syndrome.

That anaemia is a common feature of Kwashiorkor in African Kwashiorkor children has been noted by TROWELL, DAVIES & DEAN[2] but it is suggested that helminths are the cause. Many years ago NORMET[121] had made similar observations on malnourished Indo-Chinese children, recording a haemoglobin percentage ranging from 10—50. NORMET pointed out that most children of the labouring classes in Indo-China were known, through post-mortem examinations, to be loaded with helminthic parasites. So it is possible that there is a separate anaemia of Kwashiorkor which is curable by iron therapy. Vitamin E dietary deficiency is known to lead to depigmentation of the incisors of rats[7, 18]. MOORE & MITCHELL[19] report mean iron contents of the enamel of the upper incisor of vitamin E-deprived albino rats ranging from 0.02%— 0.04%, as compared with 0.24%—0.27% for control animals receiving 1 mg DL-γ-tocopheryl acetate per head, weekly. However, no such reports seem available on the bleaching of the teeth of Kwashiorkor children as a result of dietary deficiency in vitamin E.

Vitamin D shortage does not appear to be an important factor in the etiology of Kwashiorkor in deprived children. Attempts to cure the disease by the administration of vitamin D had proved deleterious in some cases[20, 21] although some paediatricians do not completely rule out the advisability of vitamin D therapy[22].

According to TROWELL, DAVIES & DEAN[2] rickets is unknown among Uganda children and it is deemed unlikely that vitamin D deficiency is one of the prime causes of the stunting of weanlings who suffer from Kwashiorkor. But they admit that in countries like South Africa where vitamin D deficiency is reflected clinically, rickets may complicate the character of Kwashiorkor. JELLIFFE[23] maintains that clinical rickets is fairly widespread in the Tropics and not uncommon in West Nigeria where irradiation of the dehydro-ergosterol of the skin is prevented by over-clothing.

Recently JONES & DEAN[26] published some interesting radiological findings on the development of the bones of the hand of children suffering from Kwashiorkor, and that of control African children. The 53 sick children were aged between 11 and 30 months. If the presence of transverse lines running across the distal end of the radius parallel to the epiphyseal margin were regarded as evidence of a past disturbance of growth, the results of these workers confirm the theory that the acute phase of the disease, Kwashiorkor, is usually imposed upon a state of chronic ill-health. Apart from the suggestion of considerable retardation of growth in Kwashiorkor, the alkaline phosphatase of blood plasma has been shown by SCHWARTZ[27] to be very low at the peak of the disease. Activation by Mg ions, and inhibition with Cu-ions, of this enzyme, are also abnormal. As treatment progresses, the level of alkaline phosphatase at first falls, then rises slowly to normal values. At the same time, the response of the enzyme to Mg^{++} and Cu^+ ions becomes typical of normal plasma alkaline phosphatase. One interesting feature of SCHWARTZ's observation is that, on recovery from Kwashiorkor, the level of this enzyme may sometimes rise to values which are diagnostic of rickets. But DEAN[26] has not been able to record any parallel radiological changes in the bones of the hand. Nor were data given by SCHWARTZ on the amounts of dietary-calcium or vitamin D upon which the subjects of the experiments were raised.

The importance of vitamin K either in the etiology or the treatment of Kwashiorkor is still largely a matter of conjecture and does not merit a review here.

ACHAR has reported[35] that about forty per cent of the cases of Kwashiorkor examined by him in India exhibited signs which were suggestive of riboflavin deficiency.

The intestinal disorders which generally accompany the onset of the weaning syndrome in children in the Tropics make it well-nigh impossible to separate what is primary avitaminosis B from that which is the indirect result of poor absorption and reduced bacterial synthesis in the gut. TURNBULL's[28] observations well illustrate the point. The Mexican children whose illness he studied were one or two years old. They showed positive liver function tests which were indicative of parenchymatous lesions. In addition, TURNBULL described the presence of glossitis, stomatitis and dermatitis which he attributed to nicotinic acid deficiency. But this clearly could have been due to a disturbance in tryptophane metabolism, or to a dietary deficiency of the amino-acid. The cheilosis and conjunctivitis which TURNBULL found in his patients were diagnosed as the effects of riboflavin deficiency. Yet CARVALHO et al.[29] and HUGHES[30] are not convinced that riboflavin therapy alone will cure such cases. TURNBULL's patients showed beri-beri-like symptoms also; there were electro-cardiograph changes and loss of the patella reflex. These the author attributed to thiamine deficiency. Biochemical support came from the low excretion of vitamin B_1 in the urine of the affected children. The experience of CASTELLANOS et al.[122] with Kwashiorkor weanlings in Cuba supports TURNBULL's clinical observations and the conclusions which he drew therefrom.

DE LA TORRE, BERKER & DURAN[31] failed to correct, by thiamine therapy, the abnormalities in the T wave and Q R S electrocardiograph complex which they found in Mexican children suffering from Kwashiorkor. In this, they are supported by GILLMAN & GILLMAN[20], TROWELL, DAVIES & DEAN[2] and many other workers. However, FRONTALLI[32] gave potassium salts to some Italian children who were ill with "distrofia da farine", and thus restored their altered electrocardiograms to normal. GOPALAN, SRIKANTIA & VENKA-TACHALAN[33] found bradycardia in four out of ten Kwashiorkor Indian children and tachycardia in three of the same group. This suggests that the observed prolongation of the Q T interval, relative to the cardiac cycle, and the reduction of amplitude of all the deflections in all the leads are not directly related to aneurin deficiency in deprived infants. The disappearance of the sinus brady-cardia on treatment of the children with milk powder suggests that the function of the myocardium may be intimately bound up with protein metabolism, especially in border-line circumstances.

CHAVARRIA[34] had previously reported the presence of tachy-cardia in his Kwashiorkor children, but here also no direct cause-and-effect relationship with vitamin B_1 intake was established.

In this connection a clue may be found in experiments such as that performed by WOHL, BRODY, SHUMAN, TURNER & BRODY[36] who found reduced co-carboxylase and thiamine concentration in the heart of patients dying of cardiac failure. In these cases, the total thiamine content of the liver was much lower than that obtained with patients without cardiac disease.

In aneurin deficiency the vitamin B_{12} content of the liver of rats has been found to be lower than normal[37]. The actual absorption of cobalamin does not only require the gastric intrinsic factor but also a so-called "intestinal B_{12} acceptor" which functions in the intestinal mucosa just as apoferritin does in iron absorption. A deficiency of the acceptor, while permitting a 90% absorption of an oral dose of 0.5 mg B_{12}, could apparently reduce to only 3% the intestinal absorption of an oral intake of 50 mg vitamin B_{12}[38].

SACHDEV[39] has shown that even on a 20% casein diet which is adequate in its provision of homocystine precursor, ligation of the pancreatic duct of rat interferes with intestinal digestion and absorption to such an extent that there is a 50% increase of liver fatty acids. This proves retractable to the subcutaneous administration of 1 mg B_{12} every other day. The total liver fatty acid averaged 4.2% for 16 Long-Evans rats which served as controls. Of the animals with ligated pancreatic ducts, those receiving vitamin B_{12} supplement had an average total liver fatty acid content of 6.18%. The mean value of the others, which were given a supplement of cobalamin, was 6.99%.

The fatty liver disease caused by protein deficiency in tropical children may be cured not only with dried skimmed milk feeds but also with synthetic diets containing banana, soya and vitamin mixture[2]. 150 g skimmed milk provides approximately 50 g protein, 1.7 g methionine, 240 mg choline and 15 mg vitamin B_{12}. 300 g soya bean paste contains roughly 50 g protein, 1.6 g methionine and 500 mg choline. The vitamin mixture added to this synthetic diet included 12 mg B_{12}.

GOMEZ, GALVAN, BIENVENU & MUNOZ[40] conducted trials on two groups of malnourished children in Mexico. To one was fed a diet containing maize, beans, wheat and soya powder. To the other was given the basic maize,beans,and soya plus an additive of milk-powder and yeast. As in the experiment of TROWELL, DAVIES & DEAN[2], both diets cured the children. However, the serum protein level took longer to return to normal in the first group than in the second. The Mexican workers did not state which fractions of the plasma protein, if any, were particularly affected.

In Kampala, THOMPSON[41] divided his Kwashiorkor children into two groups, (a) and (b). 35 subjects in group (a) were given cow's milk protein, while the 42 patients in group (b) received a preparation of soya beans with a supplement of methionine or vitamins. As criteria of the efficacy of his treatment this worker considered weight gains, loss of oedema, level of enzymes in the pancreatic juice and serum level of total proteins. Experiments in which cobalamin or methionine were omitted from the diets showed that these substances could not, per se, improve the value of soya proteins in correcting the biochemical imbalance of Kwashiorkor.

These results make it unlikely that vitamin B_{12} deficiency is an important element in the genesis of Kwashiorkor. Yet, if dietary shortage in cobalamin is not an essential feature, it will be necessary for future investigators to discover the sources of this vitamin in the nutrition of African infants whose weaning diet is practically devoid of animal proteins. Native food preparations in many parts of the tropics involve the maturation by fungal fermentation of grain or tuber, with a certain amount of microbiological synthesis of vitamin B_{12}. If the amount of cobalamin thus provided is significant, the fact that protein intake is critical in the treatment of Kwashiorkor would be explained. So, also, would the general absence of pernicious anaemia in Africa.

There is little doubt nowadays that children require B_{12} or B_{12}-like substance for adequate growth. In an experiment with eleven children whose rate of growth was much reduced, WETZEL, FARGO, SMITH & HELIKSON[42] produced with Vitamin B_{12} a spectacular increase in the growth rate of five of the children. One was cured of his allergic bronchitis.

ALBANESE, HOLT, IRBY, SNYDERMAN & LEIN[43] showed that casein, fortified with L-tryptophane and L-Cystine, was much more effective in maintaining satisfactory growth rate in three normal infants in the later half of their first year of life than was an acid

68

digest of the casein. Unfortunately, it was not established that this acid-labile growth substance was not the streptogenin shown by WOOLEY[44], [45] to be required for growth by mice.

Many animal experiments have shown that it is possible to extract growth and animal protein factors from the faeces of various species of creatures[46], [47], plants[48] and from milk[49].

The effects of folic acid and cobalamin on methyl synthesis and storage in the liver have been studied by FATTERPAKER, MARFATIA & SREENIVASAN[50] in an experiment in which 3 micrograms B_{12} was administered per mouse per day, using methionine and serine as precursors. Their results are summarized in Table XII and cast considerable light on the mode of action of dietary B_{12} on protein metabolism in the body. From these it appears that dietary cobalamin has a direct action on the synthesis of methyl groups from precursors, while folic acid may be necessary for the biosynthesis of the ethanolamine moiety of choline. The observed synergistic action of vitamin B_{12} and folic acid in trans-methylation may be explained on the supposition that cobalamin promotes methyl anabolism, e.g. methionine, from precursors, while folic acid inhibits catabolism through better utilization of formate.

Table XII

The Effect of Folic Acid and Vitamin B_{12} on methyl synthesis

Folic Acid	Vitamin B_{12}	% increase over unsupplemented	
		Choline	Methionine
—	—	8	4
—	+	15	13
+	—	5	5
+	+	23	19

Derived from a paper by FATTERPAKER, et al.[50] by courtesy of the publishers.

The results of some experiments by HENRY & KON[51] support this view. They found that in deficient rats the biological value of casein was increased to the same extent by the addition of vitamin B_{12} plus 1% homocysteine as by 1% methionine alone. With similarly prepared rats the authors observed that the presence or absence of cobalamin had no effect in altering the standard increase of biological value of casein which was brought about by 1% methionine or by a mixture of 1% homocysteine and 0.5% choline. Nor did B_{12} addition to the diet of normal rats increase the biological

(230)

value of soya or casein at the level of 16% protein intake, although there was an increase from 79.8 to 82.2 at the 8% level. This finding suggests that cobalamin is not directly connected with protein utilization, that it does not function in trans-methylation, but that it is required for the synthesis of methyl groups.

Although fresh fruits and vegetables abound in the tropics, ascorbic acid deficiency in young children is not an unknown entity. Reports have come from observers in Asia and Africa of clinical symptoms which are characteristic of vitamin C deficiency associated with Kwashiorkor, "especially in children who have been weaned at an early age to an unsuitable cereal diet"[52, 53, 54]. In some of these cases[53] the blood level of ascorbic acid was normal. And the description of petechial haemorrhages was not corroborated by the result of capillary fragility or ascorbic acid saturation tests.

The intake of ascorbic acid by a newly-weaned child depends on the kind of food to which it is transferred and on the form in which the food is served. Hence the preparation of the food prior to cooking is an important consideration. CHATTOPADHYAY & BANERJEE[55] have shown that the ascorbic acid content of Indian pulses increases greatly during the first few days of germination, while the ascorbic acid oxidase activity also enlarges. But the authors did not observe any alteration in the total ascorbic acid content from the third to the fifth day of germination. In Nyasaland and other parts of the Tropics, pulses and cereal are soaked until they germinate, if they are to be used in the preparation of the porridge to which children are weaned. PLATT[56] found that when dry, ungerminated grains are crushed into flour by machines, vitamin deficiencies manifest themselves in the children who are fed on the flour.

The habit of cooking vegetables in Asia and Africa leads to considerable losses in their ascorbic acid content. MULAY, DHOPESHWARKAR & MAGAR[57] found that the vitamin C content of raw chillies decreased by an average of 11.3% when cooked in the open air. If left exposed to the air for 30 minutes before cooking, the mean loss became 25.8%. Cooking for two minutes under 10 lb. pressure was even more destructive (Table XIII).

Although the question of the urinary excretion of 17-Ketosteroids by malnourished children is by no means settled, it is worth mentioning that RAMACHANDRAN, VENKATACHALAN & GOPALAN[58] estimated the 24 hour output of 17-Ketosteroids of 12 clinically well Indian children and found lower values than those published for European children. The mean age of the children was 5 years, and the mean daily output of the steroid 0.41 mg. Values obtained for five children with nutritional oedema syndrome were 0.78, 0.29, 0.09, 0.10, 0.20 mg/day, respectively. After 4—8 weeks treatment with high-protein diet, the output of 17-Ketosteroid increased, in

70

Table XIII

Effect of Cooking on the Ascorbic Acid content of Chillies

Vegetable	Ascorbic acid mg/100 g raw fruit	Ascorbic acid mg/100 g cooked under 10 lb. pressure
Chillies freshly cut	68.7	55.6
	145.5	123.4
Chillies exposed to air for half hour	82.8	52.3
after cutting	91.9	39.4

After MULAY, et al.[57] Quoted by courtesy of the publishers. In a similar experiment with Bantu peasants in South Africa, the mean losses of vitamin C for tests carried out on four separate days was 11% for open air cooking and 21% for pressure cooking at 12.5 lb. for 5 minutes.[59]

every case, to 1.60, 0.48, 0.15, 0.11, 0.49 mg/24 hours, respectively. Similar results were obtained with adults.

The apparent connection between vitamin C and the adrenal cortex has been the subject of much research[60], [61], [62], [63]. CLAYTON & PRUNTY[64] were unable to show any direct anti-scorbutic effect of A.C.T.H. in animals. Using the tooth-structure method for their assessment of vitamin C requirement of male guinea pigs, HARRIS, CONSTABLE, HUGHES & LOEWI[65] gave 7.5 mg cortisone daily to each of six guinea pigs receiving graded doses of ascorbic acid. Another set of six animals had identical treatment except that the cortisone was omitted. These workers found that, although cortisone had the effect of further lowering the concentration of the vitamin in the adrenal glands when the intake of ascorbic acid was low, the dose-response curves of the two sets of experimental animals were so similar as to leave little doubt of the inefficacy of cortisone in influencing the general body requirement of ascorbic acid.

Weaning Diet (High Carbohydrate — Low Protein)

The composition of the diet to which children are weaned in the under-developed communities of the Carribean, Africa and Asia follows a uniform pattern. WALKER, FLETCHER et al.[66] studied 58 samples of fluid food collected from feeding bottles and pans brought by Bantu women to the out-patients department of a hospital in South Africa, for feeding their newly weaned children who were under two years of age. A further 191 samples were collected from the homes of other children. Analysis of the protein,

(232)

ash and lipid content of the foods showed that 100% of the thin cereal paps and proprietary cereal dried milk mixtures supplied well under the 70 cal/100 ml which CLEMENTS[67] recommends for breast milk. The best thick cereal pap gave 76 cal/100 ml fluid and $2.5 \pm 1.1\%$ protein. 43 to 100% of all the other food samples studied were deficient in protein, when assessed on the basis of the recommended concentration of 2.3/100 ml artificial food[67].

The directions given on proprietary tins and packets of food for reconstitution into children's diets are sometimes so imprecise as to lead to a calorical error of 30% in some cases[66]. Even in Scotland this factor is responsible for much of the over-dilution of the proprietary brands of food given to children. Thus, HYTTEN[68] found that 78 out of a total of 100 artificial milk mixtures prepared by Aberdeen mothers for their children were deficient in calories. Many other workers in the tropics are familiar with the corn paps and mealie meal porridge to which indigenous children are weaned[69, 70], but there are very few detailed analyses of their composition.

The energy requirements of children in the age range of 4−6 years have been studied in detail by NICOL[71] in Northern Nigeria. Three groups of children were involved and the results are summarized in Table XIV. The results suggest that the greater the deviation from the standard calorie requirement the larger the mean deficiencies of weight and height of the children.

Table XIV

Comparison of Intake of Calories and Proteins of Northern Nigeria and North American children aged 4-6 years.

	Standard North American	Kanuri and Shuwa	Otukwong	Camberri
		Deviation from standard		
Number in Group	—	27	26	23
Height (ins.) [72]	43.8	— 2.2	— 3.6	— 2.4
Weight (lb) [72]	42	— 3	— 7	— 7
Intake of Calories [73]	1480	— 100	— 570	— 370
Intake of total protein (g) [74]	50	+ 28	— 24	— 1

Abridged from NICOL, B. M.[71] and published by courtesy of the publishers. The data for North American children may be compared with those given by KROGMAN[105].

The intake of protein by the Kanuri and Shuwa children was over 50% greater than the standard allowance, but there was a total calorie deficiency of 100, per child, which may have been responsible for the negative deviations of height and weight of these children.

Nitrogen balance studies on Chinese children of 4—5 years of age were conducted by KUNG & FANG[75] with similar results.

MURTHY, SARANYA REDDY, SWAMINATHAN & SUBRAHMANYAN[76] performed metabolic experiments with five Indian children of a slightly older age. All the experimental subjects had lived for two years previously on a deficient diet in an orphanage. The experimental diet conformed well to the pattern to which the children were accustomed, providing mean, daily, caloric intake of 1010. The nitrogen intake was 2.955 ± 0.53 g, derived entirely from vegetable source. 199.8 ± 3.0 mg Ca and 566.7 ± 19.0 mg P formed part of the daily consumption of the children. The experiment lasted 10 days, of which the first five were devoted to getting the children used to the diets and the conditions in which the urine and faeces were collected. The results were both interesting and surprising. All the children maintained a positive nitrogen balance, with a mean retention value of 18 mg/kg body wt. This compared well with the mean value of 20 mg/kg body wt. which was obtained by MACY[77] with six American children of the same age group, who had a mean nitrogen intake of 466 mg/kg body weight, as compared with 19/mg N/kg body wt. fed by MURTHY et al.[76]. In the Indian experiment the mean daily retentions of calcium and phosphorus were 23 mg and 62 mg, respectively. All but one of the children were in positive balance throughout, with respect to these inorganic elements. After a period of six months on the diet it was found that all five children had increased in height by nearly an inch. Three of the children put on nearly half a pound of weight, in each case. The fourth remained stationary, while the fifth actually lost $\frac{1}{4}$ lb.

These results suggest that a considerable metabolic adaptation is possible when children subsist for long periods on an ill-balanced, vegetarian diet and a low intake of calories.

On the other hand, replacing 25 or 50% of the rice resulted in diminished nitrogen and calcium retentions which were proportional to the amount of millet in the diet of seven teen-age Indian boys[126].

NICOL[71] asserts that the "caloric requirement of certain groups of Nigerian children have been under-estimated," implying that temperature, humidity, infestation with intestinal parasites, and malaria may have the combined effect of raising the basic caloric requirement. BRAY[80] found no change in N_2 balance after the elimination of intestinal parasites.

The work of MURTHY et al.[76], on the other hand, points in the direction of adaptation to a lowered calorie requirement which is, perhaps, correlated with a lowered Basal Metabolic Rate. If the

B.M.R. of Indian[78] and African children were on the average lower than the American standard, a reduced caloric intake would be required. However the margin of calorie provision which must be made over and above the B.M.R. value may vary, as between individuals and groups of individuals [79], [77].

BRAY[80] conducted nitrogen balance studies on 8 chronically malnourished West African children, aged 7—9 years, and found reduced requirements of calories and protein for positive balance, but their basal metabolism was not determined.

That profound errors of metabolism may result from feeding a high carbohydrate, low-protein diet to young children has been underlined by the work of LINDAN[81] and BASSIR[83] on growing rats. The livers of experimental animals become infiltrated with fat, and liver glycogen clearance is retarded. Essential enzymatic processes may be disrupted.

BYRON's[82] analysis of boiled plantain, a typical staple diet of tropical peoples, gave the following results:

Moisture	74.9%
Choline	0.42 mg/100 g
Inositol	45 mg/100 g
Methionine	12 mg/100 g
Nitrogen	0.23%

These figures work out as approximately 1.4 g protein per 100 g of plantain (970 cal) consumed and 0.84 g methionine is available for every 100 g plantain protein eaten. The mean deficiency in the plantain diet is the very low protein intake.

It would be interesting to know the long-term effect of feeding this kind of diet to young children, especially with regard to their calorie requirement and nitrogen balance. On account of the reported galactose tolerance in Kwashiorkor patients, ELDIN & WAFA[127] have suggested that severe cases should, at the beginning, be treated with a galactose-free high protein diet.

Inter-relationships of Vitamins, Carbohydrate and Fat

The vitamin requirement of man varies from time to time, with the diet, state of health, activity, basal metabolism, and so on. Sex, age, body size and drugs also alter the vitamin requirement of the human being.

It is probable that the requirement of a malnourished child, for vitamins of the B—group, are increased before and during the onset of a deficiency disease. It is unlikely that the need increases to the same extent in different individuals[84].

The most satisfactory biochemical method of assessing the vitamin B requirement of a child is probably the determination of the amounts of vitamin, or its metabolites, which are excreted in a

24-hour volume of urine, following the administration of known doses[85]. The extent to which vitamins B, which are synthesized by micro-organisms in the gut, are available to a malnourished child is difficult to assess. Nor does the level of a vitamin in the blood serve as a reliable pointer to the state of nutrition of the subject. The requirement of thiamine is conditioned by the number of calories utilized, the amount of carbohydrate in the diet, the fat content of the diet, the presence of antithiamine, and the action of enzymes which destroy the vitamin. No general agreement has, however, been reached as to the amount of thiamine required per 1000 calories in the normal person. The United States Recommended Dietary Allowance is 0.5 mg thiamine per 1000 calories. The work of ALEXANDER & LANDWEHR[86] pointed to a requirement of 0.44 mg/1000 cal, while HOLT[87] deduced a much lower requirement, i.e. 0.13 mg/1000 cal. There is a probable adaptation to low intakes of vitamin B_1 in man[88]. Nevertheless, malaria fever does increase the requirement of this vitamin.

Riboflavin enters into tissue oxidation processes in combination with protein. The requirement for a child is probably not more than 7.3 mg/kg body wt. per day. In protein malnutrition of weaning the requirement for this vitamin may be higher.

Niacin requirement is also influenced by the quality and quantity of dietary protein. This suggests that the deprived child would probably need more than the United States allowance of 0.17— 0.26 mg Niacin/kg body weight. The relation between the metabolism of Coenzyme I and II and pellagra is still to be shown. Niacin is a component of these Coenzymes and, therefore, enters intimately into the biochemistry of cell respiration and glycolysis. The analysis of foodstuffs and body tissues by WILLIAMS et al.[89] gave results for niacin, riboflavin and thiamine, the ratios of whose averages were 10 : 2 : 1. But there is no confirmation that these B vitamins must be present in food in this proportion for their requirements to be met.

Although megaloblastic anaemia of infancy and nutritional macrocytic anaemia both respond to folic acid therapy, it has not been established that pteroylglutamic acid is a necessary nutrient for a growing child. There is, however, a metabolic relationship between tissue amino acids and folic acid. 0.2—1.0 mg/day is the range of effective therapeutic doses.

Pyridoxal, pyridoxine and pyridoxamine are involved in the utilization of dietary protein and probably also to the causation of fatty livers in children[90]. Intestinal synthesis of these substances and their anti-metabolites, by intestinal micro-organisms, complicates the dietary requirement and makes any assessment risky.

Pantothenic acid is widely distributed in nature and is present in most common foodstuffs. Its curative value in treating the "burning feet" disease has been assessed by GOPALAN[91]. The normal requirement of this vitamin is probably about half that of niacin[89].

The anti-pernicious anaemia substance, Vitamin B_{12}, is water-soluble and is present in many natural foods. The intrinsic factor of gastric juice is required in order that vitamin B_{12} may be active. Both of these substances must be supplied in adequate quantities in the diets of the unsophisticated peoples of Africa, where pernicious anaemia is rare. There is probably no vitamin B_{12} requirement in health. But the possible influence of this vitamin on ribonucleic acid synthesis may have an important consequence in protein malnutrition in childhood.

However, the more recent observations of MACDOUGALL[123] with 18 malnourished East African children, 8 of whom had Kwashiorkor, show no relationship between the level of serum vitamin B_{12}, and hypoproteinaemia with mild nutritional anaemia.

The connection between the inter-conversion of fat and carbohydrates, with the mediation of dietary B-vitamins, has been studied in detail by HOLMAN[93] in a group of young children living in a German orphanage, with fat supplying 33% of the total calories of their experimental diets. WILLIAMS et al.[92] suggested a daily intake of aneurin of 0.45 mg/1000 cal. On a fat-free basis, their requirement value was 0.66 mg/1000 cal. This value is independent of the amount of fat in the diet. HOLMAN[93] expressed the aneurin results of his vitamin balance studies in terms of the non-fat calories provided by the diets, in relation to their energy values. This procedure is unnecessary for riboflavin.

The requirement of normal children for vitamin A depends on the criterion upon which need is based. LEWIS & HAIG[94] calculated the requirement of infants, on the basis of dark adaptation, and concluded that a daily intake of 18 to 20 I.U., per kg body weight, (i.e. about 6 mg vitamin A) is the minimal requirement. LEWIS & BODANSKY[95], on the other hand, reported that it required 100—200 I.U. vitamin A per kg body weight in order to maintain the normal blood level of 40—114 I.U. per 100 ml. It is known that, when placed on a vitamin A—free diet, children suffer a diminution of their vitamin A blood level long before dark adaptation fails[96].

Vitamin D is needed for proper growth. Therefore, the vitamin D requirement is highest in children. There is evidence that on a daily intake of 135 units of vitamin D, children do not grow as well as when given 400 units a day. On an intake of 1500 units, signs of toxicity become evident[88]. It is difficult to decide upon a precise amount as the requirement for growing children since they can sometimes obtain all the vitamin D they require by activitating the 7—dehydrocholesterol in their skin. Moreover, the requirement of vitamin D is influenced by the dietary intake of calcium and phosphorus[99].

No instance of vitamin E deficiency has been clearly established

76

in children, and the results of therapeutic administration of toco-pherol are of dubious meaning.

As with the other fat-soluble vitamins, the absorption of vitamin K is favoured by the presence of bile. All the compounds with vita-min K activity are related to 2 methyl 1 : 4 naphthoquinone. Their main physiological function lies in prothrombin formation in the liver, although the mechanisms by which the process is me-diated are unknown[98]. It is well known that the intravenous ad-ministration of vitamin K reverses the action of dicumarol in pre-venting clot formation in the blood stream. Thus, although vitamin K falls in the group of essential metabolites, it is not possible to state the exact dietary requirement.

Contradictory evidence[99] as to the requirement of lipoic acid and vitamin P makes it impossible to determine the dietary requirement of these substances for children. The fact that lipoic acid contains a thiol group and functions in connection with pyruvic acid oxi-dation, suggests that it may play a limiting role in Kwashiorkor where the normal metabolism of sulphur-containing amino acids is upset.

The biochemical role of ascorbic acid in the metabolism of dietary carbohydrate in the body is still a matter of conjecture[103, 104]. The action of vitamin C is probably related to the fact that it can effectively enter into oxidation — reduction reactions by means of the ene — diol grouping —C (OH) = C (OH). There is strong evi-dence that ascorbic acid is connected with the oxidation of the side chain of tyrosin in animal tissues[100]. Hence vitamin C action can be related to protein metabolism. Its influence on carbohydrate metabolism in plant tissues seems to be concerned with the catalytic conversion of lactate to pyruvate[101] and with changes involving triose phosphates[102]. The requirement of children with respect to ascorbic acid is unknown and various guesses have been made, in the range of 10 mg to 150 mg per day[88]. STRANSKY[52] reported the occurrence of vitamin C deficiency in Kwashiorkor, in children weaned early to a cereal diet. It is known, however, that fevers and other traumas can reduce the level at which the dynamic equili-brium (commonly called vitamin C saturation) is established[106].

Amino Acid Metabolism

The importance of amino-acids, especially lysine, methionine and cystine, in the body of a child weaned on to a high-carbohydrate low-protein diet, has been emphasized by BIGWOOD and his collab-orators[107, 108]. Cassava meal is the commonest foodstuff from which children's porridge is made in Africa and other parts of the Tropics. BIGWOOD'S[107] analysis of debittered cassava showed that there was only 3.3 mg cystine sulphur and 4.3 mg methionine sulphur per

100 g cassava. The undetermined sulphur amounted to 29.1 mg/100 g of the foodstuff. These figures compare very unfavourably with the concentrations of these fractions in barley where the corresponding figures were 63, 32 and 18 mg/100 g, respectively.

If a child were weaned at the age of two years from breast-milk, which is not deficient in the sulphur-containing amino acids, and put on a cassava gruel, the amount of the tuber required to provide the daily calorie requirement of the baby will only provide 0.024 g of sulphur from the two amino-acids, whereas 0.16 g sulphur would have been obtained from the equivalent 1.8 litres of breast-milk. These calculations do not take into account the possibility that at least a part of the undetermined sulphur of cassava may be available to the child for metabolic purposes.

It is known that human hair contains approximately 4.5% of sulphur. Its chief protein, keratin, is made up of 11—12% cystine, i.e., about 3% sulphur[107]. Since the hair and skin are among the main organs which are affected by chronic protein malnutrition of childhood, it is reasonable to presume that shortages of the sulphur-bearing amino-acids in the diet of young children would adversely affect the metabolism of hair and skin.

NAGCHAUDHURI & PLATT[109] have shown that the hair pigments of healthy African children and those suffering from malnutrition behave differently when chromatographed on paper. The authors found that pigments of similar characteristic to that of the malnourished African child could be obtained by keeping hooded rats on a low protein diet (7% casein), or on a low methionine diet (7% arachin), or on a high protein diet (16% casein). On the last diet, the rats were irradiated for long periods with ultra-violet light.

The implication of the last observation is that manifestation of the ill-effects of dietary sulphur deficiency may arise directly or indirectly through the mediation of an antagonism of one or other of the sulphur-containing amino-acids.

Hepatic injury in rats, which is due to conditioned sulphur amino-acid deficiency, has been investigated by POPPER, DE LA HUERGA & KOCH-WESER[110]. Forty-eight hours after the administration of 150 mg bromobenzene per 100 g body weight of the animal, these authors found profound biochemical changes in the liver and blood: liver esterase fell by about a third, from 200 millimol/hour/100 g to 140 millimol/hour/ 100 g. There was a two-fold increase in the level of liver alkaline phosphatase, but the total lipids of the liver were not affected. In the blood serum, the bromsulpthalein retention value rose from less than 0.3 mg/100 ml in the control to 2.5 mg. Serum bilirubin suffered a 100% increase.

The authors found that these biochemical changes could be abolished or considerably minimized by the administration of 300 mg methionine or cysteine/100 g body weight simultaneously with the

78

bromobenzene. In the latter event, the cysteine equivalent of the urinary mercapturic acid and the percentage of the bromobenzene injected were nearly doubled.

That the coupling of cysteine with bromobenzene is at the para position is indicated by the low mercapturic acid production which is obtained with the para compound, as compared with the meta. Not unexpectedly, therefore, dihalogenated benzene at the para position does not cause the biochemical lesions described above.

The effect of conditioned amino acid deficiency due to ethionine has been studied in its histological aspect in the pancreas[111, 112, 113]. Fibrositic atrophy is the usual terminal pattern obtained and this can be prevented by the simultaneous administration of methionine and ethionine. Some of the biochemical lesions which are produced in the liver by ethionine appear to be sex-linked. Thus, a single intraperitoneal injection of the drug results in fat accumulating in the livers in female rats but not in males[114]. Male animals which have previously been castrated are affected in the same way as normal female rats, while the administration of testosterone to female rats protects their livers from fatty infiltration[115].

One peculiar aspect of this conditioned amino-acid deficiency is that the concentrations of the enzymes alkaline phosphatase, esterase and succinic dehydrogenase in the liver are not affected by the administration of ethionine[110]. On the other hand, the total lipid content of female rat liver increases five-fold when the drug is administered to the animal. Methionine almost completely abolishes this effect of ethionine. The bromosulphthalein retention of the serum of female test rats increased by a factor of ten when compared with the controls. Serum bilirubin is six times as concentrated in the blood serum of ethionine-treated female rats as in those not treated.

The enzymatic lesions produced by bromobenzene in rats are similar to those observed in Kwashiorkor children[116]. And it may well be that this disease represents the sum total of a series of conditioned, rather than primary dietary deficiency.

There is growing evidence[117, 118, 119] that urea synthesis and the anabolism of protein in the body of a malnourished child are more intimately bound up than was hitherto believed, and it has been suggested that one of the effects of animal-protein malnutrition in childhood is the low blood urea and urea clearance values which are found in clinically-normal African adults[120].

From the analysis of the levels of serum protein fractions of 429 Kwashiorkor patients in Nigeria, BASSIR[124] has presented statistical data which suggest that the change during treatment in the relation of albumin to total protein may be of prognostic value in determining the efficacy of dietary treatment of Kwashiorkor.

In a similar study of 47 Egyptian children aged 6—24 months,

EL GHOLMY et al.[125] obtained a decrease to approximately 75% of normal in the serum alfa and beta globulins before treatment and a return to 97% of normal values after dietary treatment. The mean beta globulin value was 0.91 g per 100 ml after treatment, as compared with 0.83 g in normal children. Unfortunately no statistical evidence of "scatter" was presented with these results.

REFERENCES

1. BROCK, J. F. & AUTRET, M., 1952. Kwashiorkor in Africa. W. H. O. Monograph, Geneva.
2. TROWELL, H. C., DAVIES, J. N. & DEAN, R. F., 1955. Kwashiorkor. Arnold, London.
3. PENA CHAVARRIA, A. & ROTTER, W., 1938. *Rev. med. Lat. Amer.* **23**, *1027.*
4. CONESA, E. G. & CAZANAS, D., 1938. *Biol. Soc. Cubana Pediaetr.* **10**, *100.*
5. COFINO UBICO, E. & KLEE, G. A., 1942. Memoria del V Congreso Medico Centroamericano. Imprenta National, San Salvador.
6. OOMEN, H. A., 1953. *Docum. med. Geogr. trop.* **5**, *193.*
7. TROWELL, H. C., MOORE, T. & SHARMAN, I. M., 1954. *Ann. N. Y. Acad. Sci.* **57**, *615.*
8. MOORE, T. & SHARMAN, I. M., 1951. *Brit. J. Nutrit.* **5**, *119.*
9. WANG, P., GLASS, H. L., GOLDENBERG, L., STEARNS, G., KELLY, H. G. & JACKSON, R. L., 1954. *Amer. J. Dis. Child.* **87**, *659.*
10. DITLEFSEN, E. L. & STOA, K. F., 1954. *Scand. J. clin. Lab. Invest.* **3**, *210.*
11. DESSAU, F. I., LIPCHUCK, L. & KLEIN, S., 1954. *Proc. Soc. exp. Biol. Med.* **87**, *622.*
12. PIRTKIEN, R., STEEGE, H. & ARZT, G., 1954. *Z. ges. exp. Med.* **123**, *396.*
13. BUTLER, E. B. & McKNIGHT, E., 1955. *Lancet. 268.*
14. TAPPEL, A. L., 1955. *Arch. Biochem. Biophys.* **54**, *266.*
15. TAPPEL, A. L., 1954. *Arch. Biochem. Biophys.* **50**, *473.*
16. LAMBERT, H. P., SMITH, H. & RICHLEY, J., 1953. *J. infect. Dis.* **93**, *93.*
17. MOORE, T., 1943. *Biochem. J.* **37**, *112.*
18. DAM, H., GRANADOS, H. & MALTESEN, L., 1950. *Acta physiol. scand.* **21**, *124.*
19. MOORE, T. & MITCHELL, R. L., 1955. *Brit. J. Nutrit.* **9**. *174.*
20. GILLMAN, T. & GILLMAN, J., 1946. *Lancet.* **2**, *446.*
21. CLARK, M., 1951. *E. Afr. med. J.* **28**, *229.*
22. FLORES, N., 1947. *Rev. Fed. med. Guatemala,* **1**, *12.*
23. JELLIFFE, D. B., 1955. Infant Nutrition in the Subtropics and Tropics. W. H. O. Monograph, Geneva.
24. NETRASIRI, A. & NETRASIRI, C., 1955. *J. trop. Pediat.* **1**, *148.*
25. NITOWSKY, H. M., CORNBLATH, M. & GORDON, H. H., 1956. *A. M. A. J. Dis. Child.* **92**, *164.*
26. JONES, P. R. & DEAN, R. F. A., 1956. *J. trop. Paediat.* **2**, *51.*
27. SCHWARTZ, R., 1956. *J. clin. Path.* **9**, *333.*
28. TURNBULL, J. M., 1951. *Gaz. med. Mexic.* **81**, *282.*
29. CARVALHO, M., PINTO, A. G., SCHMIDT, H. M., POTSCH, N. & COSTA, N., 1945. *J. Pediat., Rio de J.* **11**, *395.*
30. HUGHES, W., 1945. *Trans. R. Soc. trop. Med. Hyg.* **39**, *437.*
31. DE LA TORRE, J., BERKER, S. & DURAN, L., 1949. *Bol. med. Hosp. Infant. Mex.* **6**, *317.*
32. FRONTALLI, G., 1953. *Boll. ed Alti dell'Accad. Med. Roma.* **61**.
33. GOPALAN, C., SRIKANTIA, S. G. & VENKATACHALAM, P. S., 1955. *Indian J. med. Res.* **43**, *15.*

80

34. CHAVARRIA, P., 1953. Proc. Macy Found. Conf. on Protein Malnutrition. Jamaica (unpublished).
35. ACHAR, S. T., 1950. *Brit. med. J.* **1**, *701.*
36. WOHL, M. G., BRODY, M., SHUMAN, C. R., TURNER, R. & BRODY, J., 1954. *J. clin. Invest.* **33**, *1580.*
37. BHAGWAT, R. V. & SOHONIE, K., 1954. *Amer. Sci.* **23**, *90.*
38. GLASS, C. B., BOYD, L. J. & STEPHANSON, L., 1954. *Proc. Soc. exp. Biol. Med.* **86**, *522.*
39. SACHDEV, J. C., 1955. *Indian J. med. Res.* **43**, *39.*
40. GOMEZ, F., GALVAN, R. R., BIENVENU, B. & MUNOZ, J. C., 1952. *Bol. med. Hosp. infant. Mex.* **3**, *543.*
41. THOMPSON, M. D., 1955. *Brit. med. J.* **2**, *1366.*
42. WETZEL, N. C., FARGO, W. C., SMITH, I. H. & HELIKSON, J., 1949. *Science* **110**, *651.*
43. ALBANESE, A. A., HOLT, L. E., IRBY, V., SNYDERMAN, S. E. & LEIN, M., 1947. *Johns-Hopk. Hosp. Bull.* **80**, *149.*
44. WOOLEY, D. W., 1945. *J. biol. Chem.* **159**, *753.*
45. WOOLEY, D. W., 1946. *J. biol. Chem.* **162**, *383.*
46. CUNHA, T. J., BURNSIDE, J. E., BUSCHMAN, D. M., GLASSCOCK, R. S., PEARSON, A. M. & SHEALEY, A. L., 1949. *Arch. Biochem.* **23**, *324.*
47. WIESE, A. C., PETERSEN, C. F. & LAMPMAN, C. E., 1948. *Poultry Sci.* **28**, *752.*
48. ZUCKER, T. F. & ZUCKER, L. M., 1950. *Vitamine und Hormone,* **8**, *1.*
49. KON, S. K., 1955. Personal Communication.
50. FATTERPAKER, P., MARFATIA, U. & SREENIVASAN, A., 1955. *Indian J. med. Res.* **43**, *343.*
51. HENRY, K. M. & KON, S. K., 1956. *Brit. J. Nutrit.* **10**, *39.*
52. STRANSKY, E., 1950. *Brit. med. J.* **1**, *1370.*
53. CHAUDHURI, K. A., 1952. *Öst. J. Kinderheilk.* **8**, *153.*
54. DIDIER, R., 1953. *Arch. franç. Pediat.* **10**, *257.*
55. CHATTOPADHYAY, H. & BANERJEE, S., 1952. *Indian J. med. Res.* **40**, *439.*
56. PLATT, B. S., 1951. Personal Communication.
57. MULAY, I., DHOPESHWARKAR, G. A. & MAGAR, N. G., 1952. *Indian J. med. Res.* **40**, *443.*
58. RAMACHANDRAN, M., VENKATACHALAN, P. S. & GOPALAN, C., 1956. *Indian J. med. Res.* **44**, *227.*
59. WALKER, A. R. P. & ARVIDSSON, U. B., 1951. *Brit. J. Nutrit.* **5**, *167.*
60. SZENT-GYÖRGYI, A., 1928. *Biochem. J.* **22**, *1387.*
61. SAYERS, G., SAYERS, M. A., LEWIS, H. L. & LONG, C. M., 1944. *Proc. Soc. exp. Biol. Med.* **55**, *238.*
62. EISENSTEIN, A. & SHANK, R. E., 1951. *Proc. Soc. exp. Biol. Med.* **78**, *619.*
63. HYMAN, G. A., RAGAN, C. & TURNER, J. C., 1950. *Proc. Soc. exp. Biol. Med.* **75**, *470.*
64. CLAYTON, B. E. & PRUNTY, F. T., 1951. *Brit. med. J.* **ii**, *927.*
65. HARRIS, L. J., CONSTABLE, B. J., HUGHES, R. F. & LOEWI, G., 1955. *Brit. J. Nutrit.* **9**, *310.*
66. WALKER, A. R. P., FLETCHER, D., STRYDON, E. S. & ANDERSON, M., 1955. *Brit. J. Nutrit.* **9**, *38.*
67. CLEMENTS, F. W., 1949. Infant Nutrition. John Wright, Bristol.
68. HYTTEN, F. E., 1954. *Proc. Nutrit. Soc.* **13**, IV.
69. WELLS, L., 1951. *Brit. J. Nutrit.* **5**, *265.*
70. WATERLOW, J. C., 1948. *M. R. C. Spec. Rep. Series, No.* **263.**
71. NICOL, BRUCE M., 1956. *Brit. J. Nutrit.* **10**, *181.*
72. STUART, H. C. & MEREDITH, M. V., 1946. *Amer. J. publ. Hlth.* **36**, *1365.*
73. FOOD AND AGRICULTURE ORGANIZATION OF THE UNITED NATIONS. COMMITTEE ON CALORIE REQUIREMENTS, 1950. F. A. D. Nutrition Stud. No. *5.*
74. NATIONAL RES. COUNCIL; FOOD AND NUTRITIONAL BOARD, 1953. Publ. Nat. Res. Counc., Washington, No. *302.*

75. KUNG, L. C. & FANG, W. Y., 1935. *Chin. J. Physiol.* **9**, *375.*
76. MURTHY, H. B., REDDY, S. K., SWAMINATHAN, M. & SUBRAHMANYAN, V., 1955. *Brit. J. Nutrit.* **9**, *203.*
77. MACY, I. G., 1952. Nutrition and Chemical Growth in Childhood, Vol. 1. Charles C. Thomas, Springfield, Ill.
78. PATWARDHAN, V. R., 1944. *Spec. Rep. Indian Res. Fund.* No. *12.*
79. YUDKIN, J., 1951. *Brit. J. Nutrit.* **5**, *177.*
80. BRAY, B., 1953. *Brit. J. Nutrit.* **7**, *3.*
81. LINDAN, O., 1943. In: Malnutrition in African Mothers, Infants and Young Children. H. M. S., London.
82. BYRON, F. E., 1955. *Proc. Nutrit. Soc.* **14**, xii.
83. BASSIR, O., 1955. *W. Afr. med. J.* **4**, *78.*
84. PETERMAN, R. A. & GOODHART, R. S., 1954. *J. clin. Nutrit.* **2**, *11.*
85. BENSON, R. A., WITZBERGER, C. M. & SLOBODY, L. B., 1953. *J. Pediat.* **23**, *437.*
86. ALEXANDER, B. & LANDWEHR, G., 1946. *J. clin. Invest.* **25**, *287.*
87. HOLT, L. E., 1944. *Fed. Proc.* **3**, *171.*
88. PETT, L. B., 1955. In: Vitamins and Hormones. Academic Press, New York.
89. WILLIAMS, R. J., EAKIN, R. E., BEERSTECHER, E. & SHIVE, W., 1950. The Biochemistry of the B-Vitamins. Reinhold Publ. Co., New York.
90. NUTRITION REVIEWS, 1954. **12**, *10.*
91. GOPALAN, C., 1946. *Indian med. Gaz.* **81**, *22.*
92. WILLIAMS, R. D., MASON, H. L. & WILDER, R. M., 1943. *J. Nutrit.* **25**, *71.*
93. HOLMAN, W. I. M., 1954. In: M. R. C. Spec. Rep. Series No. *287.*
94. LEWIS, J. M. & HAIG, C., 1939. *J. Pediat.* **15**, *812.*
95. LEWIS, J. M. & BODANSKY, O., 1943. *Proc. Soc. exp. Biol. Med.* **52**. *265.*
96. RUBIN, S. H. & DE RITTER, E., 1954. In: Vitamins and Hormones. Academic Press, New York.
97. NICOLAYSEN, R. & EEG-LARSEN, N., 1953. *Vitamins and Hormones* **11**, *29.*
98. DAM, H., 1953. *Proc. Nutrit. Soc.* **12**, *114.*
99. SCARBOROUGH, H. & BACHARACH, A. L., 1949. *Vitamins and Hormones* **7**, *1.*
100. DARBY, W. J., DE MEIO, R. H., BERNHEIM, M. L. C. & BERNHEIM, F., 1945. *J. biol. Chem.* **158**, *67.*
101. JAMES, W. D. & CRAGG, J. M., 1943. *New Phytol.* **42**, *28.*
102. JAMES, W. D., HEARD, C. R. & JAMES, G. M., 1944. *New Phytol.* **43**, *62.*
103. HARRIS, L. J., 1953. *Proc. Nutrit. Soc.* **12**, *128.*
104. MEIKLEJOHN, A. P., 1953. *Vitamins and Hormones* **11**, *61.*
105. KROGMAN, W. M., 1941. Growth of Man. *Tab. Biol.* **20**. Dr. W. Junk Publishers, The Hague.
106. KARK, R. M., 1953. *Proc. Nutrit. Soc.* **12**, *279.*
07. BIGWOOD, E. J., 1954. In: Malnutrition in African Mothers, Infants and Young Children, H. M. S. O., London.
108. BIGWOOD, E. J. & ADRIAENS, E. L., 1954. In: Malnutrition in African Mothers, Infants and Young Children. H. M. S. O., London.
109. NAGCHAUDHURI, J. & PLATT, B. S., 1954. In: Malnutrition in African Mothers, Infants and Young Children. H. M. S. O., London.
110. POPPER, H., DE LA HUERGA, & KOCH-WESER, D., 1954. *Ann. N. Y. Acad. Sci.* **57**, *936.*
111. WACHSTEIN, M. & MEISEL, E., 1951. *Proc. Soc. exp. Biol. Med.* **77**, *569.*
112. GOLDBERG, R. C. & CHAIKOFF, I. L., 1951. *Arch. Path.*, **52**, *230.*
113. DE ALMEIDA, A. L. & GROSSMANN, M. L., 1952. *Gastroenterology* **20**, *554.*
114. FARBER, E., SIMPSON, M. V. & TARVER, H., 1950. *J. biol. Chem.* **182**, *91.*
115. FARBER, E., KOCH-WESER, D. & POPPER, H., 1951. *Endocrinology* **48**, *205.*
116. WATERLOW, J. C., 1954. In: Malnutrition in African Mothers, Infants and Young Children. H. M. S. O., London.
117. KENNEY, R. A., 1953. *Brit. med. J.* **1**, *600.*
118. HOLT, L. & SNYDERMAN, S. E., 1956. In: Some Aspects of Amino Acid Metabolism, Rutgers University Press, New Brunswick.

82

119. BARNICOT, N. & SAI, F. T., 1954. *Lancet* **266**, *778*.
120. BASSIR, O., 1957. *W. Afr. J. biol. Chem.* **1**, *61*.
121. NORMET, L., 1937. *Bull. Acad. Méd., Paris* **117**, *239*.
122. CASTELLANOS, A. JNR., AZAN, L., JUNCO, J. A. & TAQUECHEL, N., 1959. *J. Pediat.* **54**, *330*.
123. MACDOUGALL, L. G., 1960. *Trans. R. Soc. trop. Med. Hyg.* **54**, *37*.
124. BASSIR, O., 1959. *Nature, Lond.* Suppl. no. **26**, *2019*.
125. EL GHOLMY, A., EL NABAWY, M., SHUKRY, A. S., ISMAIL SORAYA & EL HAWARY, M. F., 1960. *J. trop. Med. Hyg.* **63**, *137*.
126. KURIEN, P. P., NARAYAUARAO, M., SWAMINATHAN, N. & SUBRAHMANYAN, V., 1960. *Brit. J. Nutrit.* **14**, *339*.
127. ELDIN, BAD & WAFA ABDOUL, 1961. *J. trop. Med. Hyg.* **64**, *110*.

CHAPTER V

NUTRITION IN ADOLESCENCE

Those children of Tropical countries, who survive the dangers which are attendant upon being weaned from breast milk to a grossly ill-balanced diet of cereal and starch, have now to contend with the demands of rapid growth. Even in the most favourable nutritional circumstances the period of adolescence is one in which the greatest care is needed to provide such supplementation of vitamins, proteins and other essential nutrients as would prevent the onset of deficiency diseases. In Asia and Africa the shortage of animal proteins in the diets of the people and the many debilitating endemic diseases are the twin problems confronting the nutritionist which give a special significance to the biochemical aspects of tissue metabolism.

Biosynthesis of Proteins

Most of the studies on the biochemistry of tropical malnutrition, in recent years, have been confined to the period of infancy and early childhood, when the metabolic aberrations are most spectacularly displayed in Kwashiorkor. However, a few investigations of the effects of malnutrition and undernutrition in adolescent Africans have been reported. KARK[1] found that the average weight and height of 365 Bantu school girls in the age range 9—15 years were inferior to those of groups of South African European children examined. The weight discrepancy was correlated with a low mean value of haemoglobin of 11.78 g/100 ml. WINTROBE[2] considers a value of 13.4 g/100 ml as a normal haemoglobin content for European girls aged 11—15 years.

The whole question of protein synthesis, even in the adequately fed person, is still an open one which has been reviewed recently by LIPMANN[3]. The most plausible sequence would appear to be that in which the amino-acid is first activated[4, 5] and then fixed to a specific amino-acid enzyme protein[6]. Experiments in which the incorporation of radioactive amino-acids into protein is followed do indicate that the amino-acid chains are deposited on ribonucleic acid (RNA) contained in special microsomes[6, 7, 109]. LIPMANN[3] postulates that an "enzyme-mononucleotide amino acid" reacts with a nucleic acid which specifically directs the patternization of a protein.

There are indications from animal experiments that the retardation of growth, which is consequential upon a low-protein dietary

(245)

regime, is reflected in alteration of the turnover rate of ribosenucleic acid. In the liver RNA-bearing particles become less concentrated in the cytoplasm of depleted cells than in normal specimens. The amount of nucleoprotein material also decreases in animals fed on low protein diets[8]. BALFOUR[9] reared a group of immature rats on a typical West African peasant diet consisting of *Pennisetum*, ground-nuts, dried baobab leaves and salt. The mean nitrogen intake of the experimental animals was about two thirds of that of the control which were on stock diet, but the percentage retention was approximately the same in both groups of animals. Supplementation of the *Pennisetum* diet with 0.85% lysine resulted in an increase in the percentage nitrogen retention. The weights of the livers, as per cent body weight, were also less in the experimental animals. The number of nuclei, per gram of liver, was much higher in the *Pennisetum*-fed rats, indicating a decrease of cytoplasmic substances.

Biochemical investigation of the plasma proteins of adolescent children in the Congo gave total protein values ranging from 4.27 to 5.08 g/100 ml. The albumin fraction ranged from 0.27—1.32 g/100 ml. High gamma globulin values were obtained i.e. 1.79 to 2.25 g per 100 ml[10].

The lowered plasma albumin may be due to (a) failure to digest and absorb proteins from the intestines, (b) impairment of the liver in which albumin synthesis generally, but not exclusively[21-23] takes place. It is doubtful whether the lesions caused by low-protein diets and parasitic infestations, in childhood, do not continue to be felt in the period of adolescence, whatever is the nutritional status then.

In any case, reduced plasma albumin may probably be followed by a lowering of the concentrations of some enzymes[11] and other tissue proteins that are synthesized directly from plasma albumin[24, 110].

Organs like the pancreas, which are able to synthesize some of their enzymes and tissue proteins, "ab initio", from blood amino-acids, are bound to be retarded metabolically, if liver function is impaired. The elucidation of the relative importance of the protein and amino-acid precursors in the laying down of tissue proteins in growing children must await the extension of the use of the isotopic technique to human nutrition.

Recent advances in our knowledge of protein and amino acid requirements have been outlined in a fascinating paper by HARPER[105]. It has been shown that growth in rats is depressed by the addition of 0.36% of DL-threonin to an 8% casein diet, deficient in niacin and tryptophane. The rate of growth is restored to normal when the imbalance is corrected with either tryptophane or niacin[106].

One of the most intriguing problems in the extrahepatic synthesis of protein in the malnourished child is the role of coenzymes and vitamins. PLATT[11] found a reduction in acetylation of sulphonamide drugs in patients with low plasma esterase. He thinks that this may be due to shortage of co-enzyme A. Since pantothenic acid is known to be interchangeable with methionine, in the synthesis of anti-bodies, shortage of these nutrients in the diet of some tropical communities may make them even more susceptible to infection. When increased antibody formation is a necessity, in order to combat an infection, it is easy to explain the reduction in amounts of pseudo cholinesterase which has been observed[11].

VAN SANDE[12] recently prepared paper electrophoretograms of the serum proteins of two sets of Congolese who were suffering from malaria. The first group of 33 people were untreated and developed ague. The second set had been treated. It was found that the mean albumin/globulin values of the untreated people were 2.02 g/100 ml and 1.6 g/100 ml for infections with *Plasmodium vivax* and *Plasmodium falciparum*, respectively. The corresponding mean albumin values were 3.79 g/100 ml and 3.43 g/100 ml. The author quotes 1.27 ± 0.13 g globulin per cent, and 4.69 ± 0.21 g albumin per cent as the normal values for healthy Europeans living in the same area as the experimental subjects.

PLATT & GRANT[11] studied the effect of administration of Animal Protein Factor (probably containing cyanocobalamin and an anti-biotic) to adolescent Gambian children in the period of the "Puberty Spurt", since (a) a similar experiment in the South of England had yielded negative results with both normal[13] and underweight children[14] and (b) BRODY[15] regards the acceleration of growth at the onset of puberty as a sign of malnutrition. After 10 months no significant difference could be found in the growth rate of the test children who received A.P.F., and the controls who did not. The growth curve of both children rose above the standard L.C.C. (1949) normal growth curve for children aged 13—15 years. Unfortunately, the nitrogen balances and serum proteins of the children were no studied during the period of the experiment.

Nitrogen Balance in Malaria and other Endemic Diseases

The immensity of the problem of malaria in tropical countries can be seen from the results of a survey of the incidence of *Plasmodium* infection, and of splenomegaly, in a group of some 500 inhabitants of a typical savannah village in Ghana[16]. The results show an almost 100% parasite rate from the age of two years until the 14th birthday is reached. After that, there is a marked fall in the incidence of the parasite in the blood of the adolescent African. On the other hand, the spleen rate falls gradually from about 75% at the age of four years to less than 10% in the twentieth year. There was very little difference in the parasite rates of the rainy and dry seasons, but the dry season spleen rate was lower, in the adolescent years, than that of the rainy season.

There can be little doubt that the high malaria parasite rates found in growing children in West Africa must cast a heavy metabolic burden on the subjects concerned, and so modify considerably their dietary nitrogen requirements[17].

Professor PLATT and his colleagues have demonstrated, by means of field experiments in Keneba[18, 19], that supplementation of the diet of school children with milk and other proteins had the same effect as the prophylactic use of anti-malarials, namely favouring a positive nitrogen balance and increased growth. The parasitaemia of malaria was reduced when the dietary supply was good, and it increased in the "hunger season" and in conditions of nutritional stress, e.g., pregnancy in the late teen-ages[20].

The malaria survey of McGREGOR & SMITH[17] in the Gambia showed that in the adolescent years the percentage of the group of boys who were parasitized lay between 50 and 60, corresponding to a mean percentage haemoglobin of 10—11 g. When the parasite rate fell in later life to an average of about 20% of the population, the haemoglobin value rose to 12 g/100 ml blood. This inverse relationship between parasitaemia and haemoglobin formation was also observed in members of the female sex, although the general level of haemoglobin density was lower in girls than in boys. And it again emphasizes the inter-relationship between malaria, nutrition and immunity.

Similar results have been obtained by COBBAN[107] who determined the malaria parasite rate and haemoglobin concentration in the blood of 1700 young adult Nigerians. There was 6.4% symptom less parasite rate and the parasite rate of half of those complaining of fever was 17.5%. Their mean haemoglobin level was 83.0% of normal, but the general mean was 83.6%. For the 254 resident student nurses examined the mean percentage haemoglobin was 82.8 but there was less variation in their individual values.

BRUCE-CHWATT's observations on Lagos children[25] and his experience with rats and mice[36] support the postulate of "an inherited transient passive immunity against malaria" which is probably identifiable with the anti-parasitic immunity of ZIEMANN[26] and TALIAFERRO[27].

Whether this immunity is transmitted from mother to child through placental blood before birth, or through colostrum later, as in rats[37], is still an open question. Both avenues are known to be employed for the transmission of immunity to other diseases. However, there is good indication[25, 28] that the effect of this inherited immunity is worn out by the end of the sixth month of a child's life.

By the time adolescence is reached, ZIEMANN's[26] antitoxic immunity is that which is most operative. As has already been indicated above, the effectiveness of the body to combat malarial infection and reduce parasitaemia is almost doubled in the late teen-age years when tissue proteins generally are being elaborated probably at a maximal rate.

Assuming that the findings of TERZIAN, STAHLER & IRREVERRE[29] on the biochemical mechanism of the immunity of mosquitoes to Malaria is relevant here, it may be deduced that the production of specific physiological or metabolic antitoxic factors is associated with general protein anabolism, although it may not depend upon simple quantitative changes in the concentration of nitrogenous substances in the body.

While plasma, haemoglobin, and plasma and haemoglobin combined had the effect of greatly reducing the cyst numbers in two-weeks old mosquitoes infected with *Plasmodium gallinaceum*, the administration of 1% glycine, 0.1% phenylalanine, and 0.25% arginine did not depress or raise the immune response of the animals, over a four week test period.

It is a general belief that antibody is an altered serum globulin which reacts specifically to antigen but is otherwise indistinguishable from other serum globulin fractions of blood. Yet it has been shown[30, 31] that guinea pigs do not produce anti-rabbit immune globulin when injected with normal rabbit serum globulin, but respond well to treatment with aggregates containing rabbit antibody.

The gamma globulin fraction of the serum protein of youth living in parts of the Tropics where malaria is endemic is higher than normal, when judged by European and North American standards[12, 69]. The M.R.C. unit in Gambia obtained a 50% increase in γ globulin values in a malarious group of children over that protected with anti-malarial drugs[36]. Although immunological researches[32, 33] with rabbits strongly support the view that immune globulins are lodged in the γ globulin fraction of serum, this may not be true in every case[34, 35].

The eradication of malaria from island populations, such as Cyprus, may present great opportunities for investigations bearing on immunogenesis. By comparing the blood serum protein electrophoresis picture of children born in the last few years with that obtained from people of the last generation, a clue may be found. In places where non-malaria parasitic infestations are endemic, the value of immuno-electrophoretic studies may be considerably reduced, unless there exists a highly selective immunization mechanism involving the various globulin fractions.

The relationship between the sickling of erythrocytes and malaria is one to which increasing attention is being directed. LEHMAN, MOURANT & IKIN[36] examined the distribution of sickle cell haemoglobin in the Mediterranean countries and found that it was restricted to certain pockets where there was both a high incidence of malaria and miscegenation with people of African race. It has been known for a long time that these are two conditions in which sickle cells are found in the body. The one is made manifest by the presence of anaemia, while the other has no overt clinical manifestation[38]. In sickle cell anaemia, the erythrocytes have an abnormal sickle cell haemoglobin S. Mixtures of haemoglobin S are found in the cells of persons showing the sickle cell trait[39] and these can be resolved by electrophoresis[51] in phosphate buffer, at ph 6.4 and ionic strength 0.04. Genes for the synthesis of haemoglobin C and D have been shown to inter-act with the sickle-cell gene to produce diseases which are similar to sickle cell anaemia[40]. Similar interactions between the genes for the synthesis of haemoglobin E, F, G, H, I, J, K, M possibly occur in certain circumstances of which the outcome is specific diseases[41]. The physical chemistry of the process of sickling has been exhaustively investigated by ALLISON[42, 43]. It seems established that the conditions in which sickling occurs in the body are characterized by the loss of oxygen from the molecules of the sickle cell haemoglobin, and their combination to form helices. These are drawn together spontaneously to form rod-like masses of haemoglobin; thus the cell becomes distorted. Aggregates of haemoglobin C or normal haemoglobin can replace every fourth molecule of haemoglobin S in the helices, yielding different configurations. Low concentrations of urea and p-chloromercuribenzoate will prevent aggregate formation of the sickle cell haemoglobins. This probably suggests that aggregation is by hydrogen bonding in the neighbourhood of sulphydryl groups. The behaviour of destromatized pure haemoglobin solutions in converting to the para crystalline phase depends on the degree and duration of its deoxygenation. And this in turn depends on the oxygen tension[44]. Erythrocytes from sickle-cell carriers become sickled under similar conditions[42]. There is a significantly higher incidence of people with the sickle cell trait among the adult population of West Africa than among infants[36]. This gives a selective advantage of about 25%. The selection occurs between the third month of life and adulthood when 80% of the sickle cell homozygotes and 60% of the sickle cell-haemoglobin C homozygotes die.

There is a sharp division of opinion among workers on the connection between malaria and the incidence of the sickle-cell trait[111]. Work on young children and on partially or non-immune teen-agers indicate lower parasitic rates in those with the sickle-cell trait[45-47]. On the other hand, ARCHIBALD & BRUCE-CHWATT[48] and MOORE et al.[49] failed to demonstrate that sickle-cell heterozygotes are resistant to malaria, and so contradict ALLISON's observation on young children and subclinical malaria[50].

Enzyme Secretion

The reduction of serum cholinesterase in children suffering from under-nutrition has been described by WATERLOW[52] and consideration has been given to this finding in the last chapter. HUTCHINSON, McCANCE & WIDDOWSON[53] also obtained low serum cholinesterase values in conditions of semi-starvation. However, the significance of this group of enzymes is not clear even in normal metabolism and may merely be a metabolic expression of protein deficiency generally.

MUKHERJEE[54] found decreased serum cholinesterase in Indian babies with Kwashiorkor and in older persons with nutritional oedema. But the liver function tests were within the normal range. The author found a "fair amount of correlation" between the rates of increase of total serum proteins and of cholinesterase when the subjects were gradually placed on a high-protein, high calorie diet.

MUKHERJEE & WERNER[55] had shown earlier that a shortage of precursor proteins was the limiting factor for the formation of serum amylase in nutritional oedema. There appeared to be an adaptive behaviour of the enzymes to a high-carbohydrate, low-protein dietary regime.

The levels of serum alkaline phosphatase which have been found in semi-starvation are lower than the accepted normals[54, 56]. In nutritional oedema there was a progressive rise in the alkaline phosphatase value, during recovery, from 1.3 ± 0.8 Bodansky units to 5.6 ± 1.1 units. The mean of the results obtained with 17 Kwashiorkor children was 4.9 ± 0.94 rising to 9.8 Bodansky units. Similar results have been obtained by BODANSKY & JAFFE[57] with fasting rats and guinea pigs. Their results were not confirmed by those of HOUGH & FREEMAN[58] who fed protein-free diets to dogs. The livers of the dogs with whom the latter authors worked may have been damaged. The reduction of the orthophosphate-hydrolyzing enzyme, in conditions of chronic malnutrition, suggests that the concommitant reduction of total serum proteins and serum albumin is due to a deficiency of protein precursors, rather than a diminishing of the synthetic function of the liver, "per se". On the assumption (a) that urinary acid phosphatase is mainly derived from the prostate gland and (b) that the prostate is vestigial in the female but very sensitive to androgenic stimulation in the male, CLARK, BECK & THOMPSON[59] determined the concentration of acid phosphatase in 24-hr. urine samples of 153 girls and boys in the age group 10—18 years. At the same time, the amounts of creatinine and 17-Ketosteroids excreted in the urine were measured. The enzyme was assessed in terms of minute output and also in terms of number of units per mg cretinine. By means of this second device it was thought that any irregularities or inaccuracies in the timing of the collection of urine would be nullified. The results showed that only a general relationship (correlation coefficient = 0.48) could be established between the 24-hour output of 17-Ketosteroids and the number of units of acid phosphatase excreted per minute. The graph relating age with urinary acid phosphatase concentration indicated, as might have been expected, a marked sex difference. In the girls there was only a 100% increase in the rate of excretion of the enzyme, from the age of ten years to the age of 18. This compares with a 400% rise in the acid phosphatase excretion of the boys. This represents a sharp upward gradient from what is probably the onset of puberty, at the age of 12 years, to sexual maturity at about the 17th year.

The validity of the assumption that 17-Ketosteroid excretion in the urine is a measure of the amount of circulating androgen is open to questioning in the case of malnourished African youth. BARNICOT & WOLFFSON[60] have shown that the daily 17-Ketosteroid output of young Nigerians may be as low as one-half of the values found in Europeans of comparable age. And it was found impossible to correct the discrepancy by keeping Africans in Britain on a (presumably) well-balanced diet for a few years.

In view of the relationship between liver disease and sex hormone imbalance[61, 62] and the frequent occurrence of clinical symptoms of disturbance of sex hormone metabolism in African boys and girls[63], it would be worthwhile to conduct biochemical investigations to find out whether this is another legacy accruing from malnutrition in childhood or early infancy. It is unlikely that the feminization and gynaecomastia which is found in otherwise normal East African boys is an evidence of the inability of the liver to metabolize or detoxicate oestrogens. It may well be a perversion of the normal pathway of the synthesis of androgens. Obviously, much light could be shed upon this intricate problem if serial determinations of 17-Ketosteroid and acid phosphatase were made on growing African boys and girls during the years of adolescence.

Changes in the intestinal enzymes have been recorded in the protein malnutrition of early childhood[64, 65]. Amylase, lipase and trypsin are all reduced in Kwashiorkor, and are raised to higher values when the children are put on an adequately balanced diet[66]. VEGHELYI[67] found reduction in the pancreatic enzymes long before liver function and the serum proteins are affected. And he, therefore, regarded changes in the level of pancreatic enzymes as a sensitive test for protein deficiency. Similar biochemical findings were made by THOMPSON & TROWELL[68] in East Africa.

Hormones and Diet

McGREGOR & SMITH[17] made a survey of the incidence of hepatomegaly in Gambian children and found the highest incidence (50%)

in the second year of life. The incidence decreased to 12% in the 10 to 16 age range. This distribution curve ran parallel with that of malaria parasitization and splenomegaly. WALTERS & WATER-LOW[70] have made a detailed study of the etiology, incidence and histological features of fibrosis of the liver in Gambian children. As a result, these authors postulate a dual etiology of the diffuse fibrosis which was commonly found in Gambian children from poor homes. The effect of malnutrition over a long period causes damage to the liver. This damage could be easily reversible with improved nutrition, unless the liver had been previously sensitized by malaria parasitization or a similar stimulus. In the latter event, proliferation of the connective tissues leading to hepatic fibrosis would result. HIMSWORTH and his colleagues[71, 72, 73] showed that in rats two sequences of the result of dietary deprivation can be seen. The first proceeds from fatty infiltration of the liver to diffuse fibrosis; the second begins with massive necrosis and goes on to post necrotic scarring. It is possible, therefore, that the fibrosis which is common in the Tropics is the result of severe and prolonged fatty degeneration[74, 75], although the time sequence between the two conditions is consecutive rather than concurrent[64, 76, 85].

A number of workers, including NAFTALIN[77] and BEVERIDGE[78], have found that male rats succumb more easily to the ill-effects of a low protein-high carbohydrate diet than female. Although both develop actue liver necrosis, the survival time is longer in the females (79 days) than in the males (58 days)[78]. The effect of methionine dietary deficiency and the possible mediation of the sex hormones on the development of hepatic necrosis was indicated by WEICHSELBAREM[80] many years ago. A significant advance was made in this matter by FERRET[81] who found that the implantation of 5 mg oestradiol in the spleen of ovariectomized animals caused continuous oestrus if the animals were kept on a low protein-high carbohydrate diet which was designed to cause liver necrosis. Supplementation of the diet with additional cystine reversed the gynaecological picture by increasing the rate of inactivation of the oestrogen. But the addition of alfatocopherol to the necrogenic diet did not restore the ability to inactivate oestradiol in the experimental animals.

SCHWARZ's[82] investigation with rats fed on a necrogenic diet low in cystine, methionine and alphatocopherol has elucidated, somewhat, the role of the hormones of the adrenal cortex in the development of massive dietary necrosis of the liver. He found that cortisone significantly, and to equal extent in both male and female, increased the survival time of the rats. However, this ameliorating effect of cortisone did not reduce the mortality due to dietary necrosis.

Although the work of HANDLER & FOLLIS[83] and BEVERIDGE[78] demonstrates a marked effect of the addition of thyroid tissue to

92

necrogenic diets fed to rats, their interpretation of the results of their experiments may not be correct. As SCHWARZ[84] has shown, when experimental and control rats are pair-fed, thyroidectomy does not reduce the necrogenicity of a basal, deficient diet. So, it may well be that increased or decreased food intake may be responsible for the apparent thyroid action in the pathogenesis of dietary liver necrosis. This view point is strengthened by considering that (a) the beneficial effect of 0.05% propylthiouracil dietary supplement may be due to its contribution of protective sulphydryl groups and not necessarily to the antithyroid action of the drug and (b) with a diet of 8% "vitamin-free" casein, 3% brewer's yeast, 7% lard, 2% cod liver oil, 76% sucrose, 4% McCollum's salt mixture and B vitamins, but without adding alphatocopherol to the diet, NAFTA-LIN[77] showed that a moderate restriction of food intake considerably lowered the incidence of acute liver necrosis in young male rats, although a severe restriction resulted in death by inanition.

In an attempt to find out whether the pancreas exerts any hormonal influence on the development of dietary liver necrosis in rats, BEVERIDGE[78] took 32 male rats of about 140 grams body weight, and depancreatized and splenectomized ten of them. Eleven of the animals were only splenectomized. The remainder served as normal controls. All the rats were fed on a basal necrogenic diet for 290 days. The results (Table XV) showed no evidence of an influential role of pancreatic hormone on the genesis of dietary liver necrosis.

SCHWARZ's[84] experiments showed that whereas a 68% protection of experimental rats from the effect of dietary liver necrosis was obtained with an 0.2% cystine supplement, only 12% protection

Table XV

Effect of Splenectomy and Pancreatectomy on the Development of Dietary Liver Necrosis.

Groups of Experimental Animals	Number of rats used	Number of rats with necrosis	Average length time in which necrotic animals fed on the diet (days)
Normal controls.	11	1	68
Depancreatized and Splenectomized	10	3	181
Splenectomized	11	3	195

Constructed from the results of BEVERIDGE[78]

was obtained with an 0.245% L-methionine dietary supplement. The protection was even less (16.5%) with 0.5% cysteine addition.

This strongly suggests that a derangement of the oxidative function of cystine is one of the major causes of liver necrosis. Vitamin E deficiency is probably related to this oxidation-reduction mechanism. Clarification of the metabolic relationship with SCHWARZ's[84] Factor 3 will have to await a more detailed biochemical characterization of the latter substance. The extent to which the above results, obtained with experimental animals, are applicable to young Africans and Asiatics, in the unravelling of the complex biochemical inter-relationships of hormones, diet and disease, cannot be known for certain. It is interesting to note that reduced function and abnormality of structure of the thyroid[79, 86], adrenal cortex[87, 60] and pituitary[87, 88] have been observed in many malnourished Africans. Unfortunately, practically all the explanations that have so far been adduced for these disorders are clinical opinions.

Calcium and Phosphorus

The rate of growth of adolescent boys and girls depends upon the retention of minerals and their incorporation into bone tissue.

In Western countries where milk plays a large part in human nutrition, no serious problems of growth are encountered because calcium and phosphorus occur in milk in approximately ideal proportions and in adequate amounts.

The African and Asian countries are in a different category. Dietary proteins as well as calcium are usually deficient. Firstly, the Ca/P ratio is apt to be considerably reduced[89]. Secondly, the actual amounts of the three nutrients are often less than what is considered ideal in Europe and Northern America. In a survey made of peasant diets in the Gambia, the daily intake of calcium, per person, ranged from 128 mg to 469 mg. The corresponding phosphorus values were 475 mg to 1406 mg. The Ca/P ratio was in all the cases studied less than 0.42. The heights and weights of the teenage Gambian children averaged nearly the same as children in the London County Council area in 1905—12, i.e., 15 to 20 per cent lower than the present-day English means[89].

The values of dietary phosphorus quoted above do not take into account phytic acid. However, as maize and other grains form the staple diets in Northern Nigeria and many other parts of the Tropics, a considerable proportion of the phosphorus contents of the diets does not contribute to growth.

Recent work with ^{45}Ca on the effect of food phytates in boys, having a typical United States breakfast, shows that, when the calcium content of the diet amounted to 239 mg and the phytic acid phosphorus equals 80 mg, the uptake of calcium was not signi-

94

ficantly lower than that obtained with phytate-free milk diets of the same calcium concentration[90]. It is significant that, when the values of calcium and phytate phosphorus, respectively, were 85 mg and 100 mg, the uptake of calcium in school boys was considerably reduced[91]. The phytate was provided by oatmeal, while the main source of calcium was milk.

With African children of a much younger age-group, who were weaned on to a low-protein cereal diet and who subsequently developed Kwashiorkor, JONES & DEAN[92] demonstrated decalcification and retardation of development of the bones of the hand, amounting to a chronological depression of 10 to 20 per cent of the normal. Earlier, in West Africa, WIENER & THAMBIPILLAI[93] had found that the bone ages of adolescent Ghana children were lower than the chronological ages by an average of 16 months. Using the American standard of FLORY[95] as normal, MACKAY[94] obtained similar results with East African teen-age children.

Isotopic tracer work with experimental animals, in the period just before the attainment of sexual maturity, indicate that on low calcium diets of the type referred to above, soft tissues take precedence over skeletal in satisfying the needs of growth. Thus, while tissue differentiation in the intercostal region is retarded by a diet consisting of 0.025% Ca and 0.064% P, the size of the liver and its general metabolism are not affected[96]. By raising the P/Ca ratio to $\frac{0.17\%}{0.024\%}$ the differentiation of calcified and cartilage cells returns to the normal pattern. But the body weight increase proceeds at a lower rate until both the amounts and the ratio of dietary calcium and phosphorus are satisfactory.

In the hot climate and bright sunshine which prevail in the Tropics the availability of vitamin D is no problem, except in some places like Western Nigeria where babies are over-clothed. Older children probably obtain all the vitamin D they require from solar irradiation of their skins. It would appear, also, that a high ambient temperature favours the deposition of phosphorus in bone. SEALANDER[97] kept ten rats in an incubator at $35° \pm 1.5°$ C for 30 days before injecting each of them with about 20 micro curies of ^{32}P intraperitoneally. 48 hours after the ^{32}P injection the animals were killed and the different tissues examined. More radioactive phosphorus was found in the bone, liver and adrenals of these animals than in the organs of the 5 controls which were kept for a month at a temperature of 26.5° C. 11 animals kept at $2° \pm 1.5°$ C for comparable periods fixed less ^{32}P than the controls. If this observation were applicable to mankind, it would help to explain the relative rarity of rickets in the tropics despite the low calcium content of breast milk and staple diets of some African and Asiatic peoples[98, 99].

(256)

Even so, the special need for calcium by adolescent boys and girls probably presents some difficulty in the biochemical reactions connected with the utilization of dietary calcium and phosphorus. For example, when radiocalcium, [45] Ca, was injected intravenously into adolescent boys, BROMNER, HARRIS, MALETSKOS & BENDA[100] found that the mean rate of disappearance of the mineral from the blood stream was 19% per minute. After five days, the percentage of injected calcium excreted in the faeces and urine was less than 8. This compared with a 20% excretion in the young adult.

By plotting values of calcium intakes against faecal excretion and extrapolating the curve to zero it is possible to obtain an estimate of endogenous calcium[101]. The difference of intake and excretion values for calcium can thus be corrected for endogenous calcium to give the amount of Ca absorbed. It has been shown that in an adult man, when the intake lies in the range 400—599 mg daily, the percentage absorbed was 43.34. 35% absorption was obtained with intakes of 600—999 mg. At intakes of 1000—1199, the percentage absorption fell to 28.

In view of the derangement of thyroid function which is sometimes associated with protein malnutrition in the Tropics, it is interesting to compare the metabolism of calcium in hypothyroidism and hyperthyroidism with that of persons with normal thyroid function.

KRANE et al.[102] showed that in myxoedema of long standing the subjects were in slightly negative Ca balance. In hyperthyroidism, the patients were in negative calcium and phosphorus balance to varying extents, although they were practically in nitrogen equilibrium. This suggests that the high urinary calcium content in thyrotoxicosis may not be directly related to nitrogen loss. On the other hand, both the anabolic and catabolic processes in bone are accelerated in hyperthyroidism.

This relationship between nitrogen loss and calcium retention must be seen in the light of the findings of many workers who find a more favourable calcium and phosphorus retention in experimental animals fed on a high protein diet, than in those having protein-deficient diets[98, 103, 104]. More recent work with [45]Ca and [85]Sr in rats has suggested that the absorption of calcium and strontium in the gut may be enhanced by lysine[108] but the evidence is rather slender.

96

REFERENCES

1. KARK, E., 1953. *S. Afr. J. med. Sci.* **18**, *190.*
2. WINTROBE, M. M., 1946. Clinical Haematology. Lea & Febiger, Philadelphia.
3. LIPMANN, F., 1956. Currents in Biochemical Research. Interscience, London.
4. SPIEGELMAN, S., HALVORSON, H. O. & BEN-ISHAI, 1955. In: Amino Acid Metabolism. Johns Hopkins Press, Baltimore.
5. ZAMECRICK, P. C. & KELLER, E. B., 1954. *J. biol. Chem.* **209**, *337.*
6. HOAGLAND, M. B., 1955. *Biochem. Biophys. Acta* **16**, *288.*
7. LITTLEFIELD, J., KELLER, E. B., GROSS, J. & ZAMECRICK, P. C., 1955. *J. biol. Chem.* **217**, *111.*
8. BASSIR, O., 1955. *W. Afr. med. J.* **4**, *78.*
9. BALFOUR, BRIGID, 1954. In: Malnutrition in African Mothers, Infants and Young Children. H. M. S. C., London.
10. DRICOT, C., BEHEYT, P., & CHARLES, P., 1951. *Ann. Soc. belge Méd. trop.* **31**, *581.*
11. PLATT, B. S. & GRANT, M. W., 1954. In: Malnutrition in African Mothers, Infants & Young Children. H. M. S. O., London.
12. VAN SANDE, MARC, 1956. *Ann. Soc. belge Méd. trop.* **36**, *335.*
13. BENJAMIN, B. & PIRRIE, G. D., 1952. *Med. Offr.* **87**, *137.*
14. MIDDLESEX COUNTY COUNCIL, 1951. Administration of Vitamin B_{12} to underweight children. – Unpublished.
15. BRODY, S., 1945. Bioenergetics of Growth. Reinhold Publ. Co., New York.
16. COLBOURNE, M. J. & WRIGHT, F. B., 1955. *W. Afr. med. J.* **4**, *161.*
17. McGREGOR, I. A. & SMITH, D. A., 1952. *Trans. R. Soc. trop. Med. Hyg.* **46**, *403.*
18. HASLET, A. W., 1952. *Sci. News,* **23**, *108.*
19. HASWELL, M. R., 1953. *Colon, Res. Studies,* No. **8**. H. M. S. O.
20. PLATT, B. S., 1952. Personal Communication.
21. BENHOLD, H., KYLIN, E. & ST. RUOZNYAK, 1938. In: Die Eiweisskörper des Blutplasmas. Dresden.
22. WHITE, A. & DOUGHERTY, T. F., 1946. *Ann. N. Y. Acad. Sci.* **46**, *859.*
23. TARVER, H. & REINHART, W. D., 1947. *J. biol. Chem.* **167**, *395.*
24. COHEN, S., 1954. *S. Afr. J. med. Sci.* **19**, *54.*
25. BRUCE-CHWATT, L., 1952. *Ann. trop. Med. Parasit.* **46**, *173.*
26. ZIEMAN, H., 1924. In: Handbuch der Tropenkrankheiten. Leipzig.
27. TALIAFERRO, W. H., 1941. In: A Symposium on Human Malaria. Amer. Assoc. Adv. Science, New York.
28. COLBOURNE, M. J. & SOWAN, E. M., 1956. *Trans. R. Soc. trop. Med. Hyg.* **50**, *82.*
29. TERZIAN, L., STAHLER, N. & IRREVERRE, F., 1956. *J. Immunol.* **76**, *308.*
30. ADLER, F. L., 1956. *J. Immunol.* **76**, *217.*
31. VAN DER ENDE, M., 1940. *J. Hyg.* **40**, *377.*
32. TISELIUS, A. & KABAT, E. A., 1939. *J. exp. Med.* **69**, *119.*
33. VAN DER SCHEER, J., BOHNEL, E., CLARKE, F. H. & WYCKOFF, R. W., 1942. *J. Immunol.* **44**, *165.*
34. LAURELL, A. B., 1955. *Acta path. microbiol. scand.* Suppl. No. **103**.
35. PETER, H., HAUSER, A. & VON DER EMDEN, A. Z., 1954. *Immunitätsforsch.* **111**, *44.*
36. COL. MED. RES. COM. XIth Annual Report, 1956.
37. TERRY, R. J., 1956. *Trans. R. Soc. trop. Med. Hyg.* **50**, *41.*
38. TALIAFERRO, L. W. M. & HOCK, J. G., 1923. *Genetics* **8**, *594.*
39. POMBURG, TANO, H. A., SINGER, S. J. & WELLS, I. L., 1949. *Science* **110**, *543.*
40. ITANO, H. A., 1953. *Science* **117**, *89.*
41. ALLISON, A. C., 1956. *Trans. R. trop. Med. Hyg.* **50**, *185.*
42. ALLISON, A. C., 1954. *Trans. R. Soc. trop. Med. Hyg.* **48**, *312.*
43. ALLISON, A. C., 1954. *Ann. Eugen., Camb.* **19**, *39.*

44. LANGE, R. D., MINNICH, V. & MOORE, C. V., 1950. *J. Lab. clin. Med.* **36**, *848.*
45. EDINGTON, G. M., 1954. *Brit. med. J.* **1**, *1189.*
46. DELIYANNIS, G. & TAVLARAKIS, N., 1955. *Brit. med. J.* **2**, *301.*
47. RAPER, A. B., 1955. *Brit. med. J.* **1**, *1187.*
48. ARCHIBALD, H. M. & BRUCE-CHWATT, L., 1955. *Brit. med. J.* **1**, *970.*
49. MOORE, R. A., BRASS, W. & FOY, H., 1954. *Brit. med. J.* **2**, *630.*
50. ALLISON, A. C., 1954. *Brit. med. J.* **1**, *290.*
51. SHOOTER, E. M. & SKINNER, E. R., 1956. *Trans. R. Soc. trop. Med. Hyg.* **50**, *201.*
52. WATERLOW, J. C., 1950. *Lancet* **i**, *908.*
53. HUTCHINSON, A. O., MCCANCE, R. A. & WIDDOWSON, E. M., 1951. M. R. C. Spec. Rep. Series. No. **275**
54. MUKHERJEE, K. L., 1957. *Bull. Cal. School. trop. Med.* **5**, *13.*
55. MUKHERJEE, K. L. & WERNER, G., 1954. *J. Lab. clin. Med.* **43**, *727.*
56. SCRIMSHAW, N. S., 1955. Protein Malnutrition. Camb. Univ. Press, Cambridge.
57. BODANSKY, A. & JAFFE, H. L., 1931. *Proc. Soc. exp. Biol. Med.* **29**, *197.*
58. HOUGH, V. H. & FREEMAN, S., 1942. *Amer. J. Physiol.* **138**, *184.*
59. CLARK, L. C., BECK, E. & THOMPSON, H., 1951. *J. clin. Endocrinol.* **11**, *84.*
60. BARNICOT, N. & WOLFFSON, D., 1952. *Lancet* **1**, *893.*
61. LONG, R. S. & SIMMONDS, E. E., 1951. *Arch. intern. Med.* **88**, *762.*
62. BISKIND, M. S., 1949. *Rev. Castro.* **16**, *220.*
63. DAVIES, J. N., 1949. *Brit. med. J.* **2**, *676.*
64. CARVALHO, M., SCHMIDT, M. M. & PINTO, A. G., 1947. *J. Pediat.,Rio de J.* **13**, *141.*
65. VEGHELHYI, P. V., 1948. *Lancet* **1**, *497.*
66. DEAN, R. F. & SCHWARTZ, R., 1953. *Brit. J. Nutrit.* **7**, *131.*
67. VEGHELYI, P. V., 1948. *Acta paediat., Stockh.* **36**, *128.*
68. THOMPSON, M. D. & TROWELL, H. C., 1952. *Lancet* **1**, *1031.*
69. EDOZIEN, J. C., BOYO, A. & MORLEY, D. C., 1960. *J. clin. Path.* **13**, *118.*
70. WALTERS, J. H. & WATERLOW, J. C., 1954. M. R. C. Spec. Rep. Series No. **285**.
71. HIMSWORTH, H. P. & GLYNN, L. E., 1944. *Clin. Sci.* **5**, *93.*
72. GLYNN, L. E., HIMSWORTH, H. P. & LINDAN, D., 1948. *Brit. J. exp. Path.* **29**, *6.*
73. HIMSWORTH, H. P., 1950. The Liver and its Diseases. B. H. Blackwell, Oxford.
74. HIMSWORTH, H. P. & GLYNN, L. E., 1944. *Lancet* **i**, *457.*
75. FERNANDO, P. B., MENONZA, O. R. & RAJASURIYA, P. K., 1948. *Lancet* **ii**, *205.*
76. DIBLE, J. M., 1951. *Brit. med. J.* **i**, *833.*
77. NAFTALIN, J. M., 1954. *Ann. N. Y. Acad. Sci.* **57**, *862 & 869.*
78. BEVERIDGE, J. M., 1954. *Ann. N. Y. Acad. Sci.* **57**, *873.*
79. GILLMAN, J. & GILBERT, C., 1954. *Ann. N. Y. Acad. Sci.* **57**, *737.*
80. WEICHSELBAREM, T. E., 1935. *Quart. J. exp. Physiol.* **25**, *363.*
81. FERRET, P., 1950. *Brit. J. exp. Path.* **31**, *590.*
82. SCHWARZ, K., 1951. *Science* **113**, *485.*
83. HANDLER, P. & FOLLIS, R. H., 1948. *J. Nutrit.* **35**, *669.*
84. SCHWARZ, K., 1954. *Ann. N. Y. Acad. Sci.* **57**, *877.*
85. GYORGY, P., STILLER, E. T. & WILLIAMSON, M. B., 1943. *Science* **98**, *518.*
86. VINT, F. W., 1949. *E. Afr. med. J.* **26**, *58.*
87. GILLMAN, J., & GILLMAN, T., 1951. Perspectives in Human Malnutrition. Grune and Stratton, New York.
88. GILLMAN, T., 1942. The Cytology of the Human (Bantu) Pituitary Gland. Thesis, Rand University.
89. BASSIR, O., 1951. The Importance in Nutrition of the Amounts of Phosphorus in poor Dietaries. Thesis, London University.
90. BROMNER, F., HARRIS, R. S., MALETSKOS, C. J. & BENDA, C. E., 1956. *J. Nutrit.* **59**, *393.*
91. BROMNER, F., HARRIS, R. S., MALETSKOS, C. J. & BENDA, C. E., 1955. *Nutrit. Abstr.* Abs. 3868. **25**.

92. JONES, P. R. & DEAN, R. F., 1956. *J. trop. Pediat.* **2**, *51.*
93. WIENER, J. S. & THAMBIPILLAI, 1952. *Amer. J. phys. Antrop.* **10**, *407.*
94. MACKAY, D. H., 1952. *Trans. R. Soc. trop. Med. Hyg.* **46**, *135.*
95. FLORY, C. D., 1936. *Monogr. Soc. Res. Child Rev,* **i**, No. **3**.
96. BASSIR, O., 1955. *J. trop. Med. Hyg.* **58**, *210.*
97. SEALANDER, J. A., 1956. *Amer. J. Physiol.* **186**, *227.*
98. BASSIR, O., 1956. *J. trop. Med. Hyg.* **59**, *139.*
99. UGA, Y., 1935. *Tohoku J. exp. Med.* **25**, *169.*
100. BROMNER, F., HARRIS, R. S., MALETSKOS, C. J. & BENDA, C. E., 1956. *J. clin. Invest.* **35**, *78.*
101. BRINE, C. L. & JOHNSTON, P. A., 1955. *J. clin. Nutrit.* **3**, *418.*
102. KRANE, S. M., BROWNELL, G. L., STANBURY, J. B. & CORRIGAN, H., 1956. *J. clin. Inverst.* **35**, *874.*
103. TARJAN, R., 1955. *Acta physiol. Hung.* **8**, *127.*
104. TARJAN, R., SZOKE, K. & SZALAY, E., 1956. *Acta physiol. Hung.* **10**, *75.*
105. HARPER, A. E., 1959. *Fed. Proc.* **18**, *104.*
106. MORRISON, MARY & HARPER, A. E., 1960. *J. Nutrit.* **71**, *296.*
107. COBBAN, Mc. L. 1960. *T. trop. Med. Hyg.* **63**, *233.*
108. RAVEN, A. M., LENGEMANN, F. W. & WASSERMAN, R. H., 1960. *J. Nutrition,* **72**, *29.*
109. SORM, F., 1961. In: *Proc. 5th Int. Congr. Biochem.,* London.
110. EDOZIEN, J. C., 1961. *Pediatrics,* **27**, *325.*

CHAPTER VI

PREGNANCY IN THE MALNOURISHED

Diet

The absence of obvious disease is not a good indication of the state of nutrition of an individual. The failure of food to nourish the body under all circumstances and conditions may not immediately reveal itself in clinical manifestations, even in changes in the biochemical processes of the body.

"Hollow hunger" is the result of underfeeding and its consequences may remain hidden for years. It is essentially a socioeconomic problem and one which is of considerable prevalence in the Tropics, particularly in the urban centres. If the degree of underfeeding is severe, then, of course, deficiency diseases occur. In "Hidden Hunger", misfeeding is the danger. Man's feeding habits are not easily changed. It is incredible that there are people in Africa whose habitual diet is deficient in essentials such as animal protein and vitamins, despite the fact that fishes, rodents and green vegetables abound in the country.

Where the diet is inadequate, the resulting malnutrition syndrome will be Primary, whereas, if the nutritional failure is conditioned by factors other than the dietary intake, the deficiency is Secondary. Conditioned nutritional deficiency may be caused by factors which alter the normal pattern of ingestion and absorption of nutrients from the intestinal tract and their utilization by the body, or it may be due to conditions which increase the body's requirement of certain nutrients, or make inefficient use of them.

There is individual variation in the characteristic physiological capacity to utilize the chemical substances which are obtained from the ingested food[1], and the biochemical fluctuation may be considerable in pregnancy and lactation[2, 3].

MACY & MACK[1] are of the opinion that the kind and amount of food taken by a pregnant woman has profound effects, not only upon her own health, but also on the course and trauma of her labour and delivery and the speed of her recovery therefrom.

Most people agree that the interrelationship between the nutritional status of women before and during pregnancy and the health of their offsprings is a close one [4, 5, 6].

McGANITY, CANNON et al.[7] studied the relationship between obstetric performance and the status of nutrition of 2046 pregnant American women of low or moderate socio-economic background. Their approximate nutrient intakes were treated statistically

(261)

(Table XVI) and compared with their foetal complications and physical and laboratory findings.

Table XVI

Nutrient Intakes of Pregnant Women, with ranges about the means or medians. *

Nutrient	— 2 S.D.	— 1 S.D.	Mean	+ 1 S.D.	+ 2 S.D.
Total Calories	1000	1500	2100	2750	3250
Protein (g)	40	50	72	90	110
Calcium (g)	0.25	0.75	1.10	1.5	2.0
Iron (mg)	6.0	10	13.5	18	22
Thiamine (mg)	0.7	1.0	1.4	1.9	2.2
Riboflavin (mg)	0.5	1.5	2.5	3.5	4.5
Niacin (mg)	4.0	8.0	12.0	16.0	20
	2.5%	16%	Median	84%	97.5%
Vitamin A. (I.U.)	1000	3000	6000	11000	19000
Ascorbic acid (mg)	20	40	60	100	160

* From a paper by McGanity et al.[7] published by courtesy of the publishers, the C.V. Mosby Company, St. Louis and the authors.

It was found that pregnancy followed an uncomplicated course in mothers whose dietary intake exceeded 1 or 2 standard deviations above the mean of the entire group. Women whose dietary intakes fell below the group mean by one standard deviation or more had a significantly higher incidence of toxaemia and other pregnancy complications. Pregnancy disease was associated with lowered intake of nutrient and total calories and also with raised total serum protein. At parturition the serum vitamin A level rose in these cases of inadequate intake of nutrients.

The relationship between maternal malnutrition in a group of 72 mothers (who gave birth prematurely) and their prematurity was not clear. But the ascorbic acid intake of the mothers was lower than the mean for the whole group. The excretion of N-methylnicotinamide and the niacin intake of eclamptic and pre-eclamptic patients were lower than expected. In their investigation of the nutritive status of pregnant women, DARBY, CANNON & KASER[8] used the assessment of dietary intake, clinical appraisal and the blood levels of haemoglobin, vitamin C, total protein, vitamin A, carotene, thiamin, riboflavin and alkaline phosphatase.

MACY & MACK[1] recorded a series of biochemical estimations of some of these blood constituents and found that their levels varied with dietary intake, socio-economic position and race. In another study MACK, KELLY & MACY[9] made dietary evaluations[10] of 46 women who suffered toxaemia during pregnancy. The medians of the nutritional ratings of the White and Negro women in the group were 64% and 53%, respectively. The medial ratings of women whose pregnancies were uncomplicated were 64% and 74%, respectively. This suggests, if anything, that Negroes require a more severe degree of malnutrition to precipitate toxaemia. Yet, there appears to be a distinct racial difference between the blood carotenoid levels of Negro and White eclamptic and pre-eclamptic women[7]. It is usual to regard 60 mg carotene per 100 ml, and less, as inimical to good health, while the normal range is taken as 100—300 mg/ 100 ml. White women with eclampsia or pre-eclampsia consistently had higher blood carotene concentrations than those whose pregnancies ran their normal course. With Negro women, the lower levels of blood carotene were found in pre-eclamptic and eclamptic women than in those with incomplicated pregnancies.

PLATT[11] refers to the excessively high toll of infant lives in African countries and involves maternal malnutrition in pregnancies with other possible causative agents. WOODRUFF has reviewed the in-increased need for iron in pregnancy[104]. His subjects with severe anaemia of pregnancy were assessed as having low protein intake[12].

The total protein of serum of Africans whose protein intake is below 50 g per day is at least as high as that of well nourished Europeans. However, the ratio of albumin to globulin in these Africans is usually less than one[13, 14]. The fraction most affected being the gamma globulin.

It is known that during the course of an uncomplicated pregnancy, alpha and beta globulins and fibrinogen increase progressively in the blood, while albumin and gamma globulin decrease in concentration[15, 16]. It therefore seems as if both the determination of total plasma proteins in the blood of pregnant Africans and the electrophoretic fractionation of these proteins are not likely to be good indices by which to detect moderate protein deficiencies resulting from the stress of pregnancy[17]. If it were to be discovered, subsequently, that there is a statistical difference between serial electrophoretograms of the plasma proteins of well nourished Africans during pregnancy and the corresponding patterns obtained with European and North American women, this last conclusion would be invalid. The result of the electrophoretic separation of plasma proteins of anaemic women in Nigeria by WOODRUFF[25] does not help because control experiments with non-pregnant Nigerian women were not described. The difficulties of comparing the results of biochemical investigations of the plasma proteins of Negro and European women in pregnancy is well illustrated by MACK, KELLY & MACY[9] who found total plasma protein levels consistently above 6.0 g per 100 ml in American Negroes with toxaemia of pregnancy, while the values for several "White" women fell below this border-line of so-called hypo-albuminaemia.

Ascorbic acid metabolism in pregnancy is of special interest in view of the involvement, with protein metabolism, in the adaptive role of the adrenal cortex. In fact, the biochemical aberrations in stress can be compared with the metabolic changes of pregnancy[18], although no direct relationship has been established between the incidence of toxaemia and the ascorbic acid concentration of blood[9].

Since this vitamin is not stored for long in the body, differences in the blood level in different individuals probably means differences in the intake of ascorbic acid. Very low (less than 0.2 mg/100 ml plasma) values have been reported in apparently normal Africans. MACK, KELLY & MACY[9] obtained results of vitamin C estimation in the blood plasma of pregnant "White" and Negro women which suggested that the levels were more indicative of socio-economic differences than genetical ones.

While it is known that ascorbic acid metabolism in the body is related to growth and the maintenance of normal cell function, it is not absolutely clear whether vitamin C has a special role in pregnancy. High concentrations of ascorbic acid have been found in the placental villi and the adrenal glands during pregnancy[9]. Low vitamin C intake in pregnancy has been incriminated with respect to the birth of mentally deficient children[36].

Anaemia of Pregnancy

Megaloblastic anaemia of pregnancy has been reported from many parts of the Tropics and a number of observers have reported that this anaemia responds well to haemopoietic substances[20-22]. NA-PIER[23] experienced difficulty in eliciting haematological response during treatment of his cases. WOODRUFF[24, 25, 26] reports similar experiences in his attempts at treating pregnant women in Nigeria who were very ill with anaemia. No improvement in the haematological picture or reticulocyte crisis was evoked by liver extract, folic acid, or vitamin B_{12}; but this was not surprising since the B_{12} concentration of the blood of 5 of the women lay in the range of 470—1000 micromicrograms per ml. MOLLIN & ROSS[88] give the normal range as 100—720 micromicrograms/ml. According to WOODRUFF[26] the type of anaemia most commonly encountered in pregnant women in Nigeria is associated with hepatomegaly and a macronormoblastic marrow. Table XVII gives a typical picture of the haematological findings.

It is obviously of interest to determine the cause of the anaemia of pregnancy in these tropical peoples, especially in view of the fact that there are so many possibilities, including malaria parasitization, infestation by worms and dietary insufficiency of proteins.

Table XVII

Haematological Findings in 40 Nigerian Women before Treatment for Anaemia of Pregnancy*.

Type of Anaemia	Haemoglobin (g/100 ml)		Red Blood Cells (millions/Cumm)		M. C. V. (cu. u.)		M. C. H. C. %		Recticulocyte (% of R.B.C.)	
	Mean	Range	Mean	Range	Mean	Range	Mean	Range	Mean	Range
Macrocytic (MCV 395 cu.g)	6.93	5.18—8.7	2.23	1.81—2.9	102.76	95.3—120.4	30.33	26.6—33.3	4.51	1.2—9.5
Normocytic (M.C.V. 82-95 cu.g.)	5.55	4.62—9.18	2.0	0.89—3.98	86.88	82.3—93.46	33.04	25.2—38.0	4.49	0.5—16.6
Microcytic (M.C.V. < 93 cu.U)	6.66	4.14—9.62	2.60	1.38—3.36	74.26	57.7—81.81	37.47	28.6—37.68	3.3	0.5—6.0

* From a paper by WOODRUFF[26], Brit. med. J. 1955, 1, 1297, quoted by courtesy of the author and publishers.

WOODRUFF[26] is inclined to lay little emphasis on the role played by the first two factors in the etiology of anaemia. In Nigeria babies born of anaemic mothers were found to be about 2 lbs (0.9 kg) less heavy than those whose parents were healthy[24]. BURKE, HARDING, & STUART[27] have shown that protein malnutrition in a pregnant woman may result in a low birth weight of the baby.

It is interesting to note that the mean dietary intake of 14 of WOODRUFF's patients[26] was as follows: total calories 1343, protein 36.4 g, fat 46.3 g, carbohydrate 195.1 g. That the demands of protein nutrition in pregnancy are a major contributory factor to the genesis of anaemia is supported by the fact that, in Nigeria, the frequency of twin pregnancies (and their concomitant heavy nutritional taxation) was four times as great in anaemic mothers as in those not suffering from the disease.

During normal pregnancy there is a tendency for the concentration of haemoglobin to decline as term is gradually approached. This is mainly due to haemodilution. Many workers regard levels of below 10 g/100 ml blood as diagnostic of anaemia [28-30]. But the decline in haemoglobin concentration during normal pregnancy is seldom so severe as to bring the blood level down to 10 g/100 ml[9]. While there is a marked progressive decrease in blood haemoglobin in some pre-eclamptic and eclamptic North American women, the Negro subjects studied by MACK et al.[9] actually showed a raised concentration in some cases. This can be partly explained by the fact that there is haemoconcentration at the height of toxaemia[31].

Although a decline in the level of haemoglobin is not necessarily an index of the nutritional status of a pregnant woman, it is not accidental that the concentration of blood haemoglobin is normally less in the malnourished than in well-fed persons[9, 24, 26]. Many observers in the Tropics have reported very low haemoglobin values in the blood of poorly nourished people, and amounts ranging from 4 to 7 g per 100 ml are not uncommon[26, 32].

Attempts at finding an explanation of the slow rise in haemoglobin concentration of the blood of the affected subjects, when treated with iron compounds, liver extracts, vitamin B_{12} and blood have not yielded much success, except to point to early maturation of bone marrow cells as one of the attendant features[26]. Evidence is being gathered in the connection between the frequency of the sickle-cell trait and the incidence of anaemia of pregnancy in tropical peoples, but it is unlikely that the presence of haemoglobin C will be found to be a major contributary factor in the development of refractory anaemia in the pregnancy of Asian and African women.

Very low blood haemoglobin levels have been reported, by many workers in temperate countries, in cases of tubal pregnancy [33-36], and gross anaemia has been adjudged by some as diagnostic of this abnormal form of pregnancy [37-39]. We do not know what the incidence of tubal pregnancy is among the malnourished communities of the Tropics, although LAWSON's[40] figures suggest that it is fairly high. Ranges of 0.33—0.95 per cent of total deliveries have been reported in the United States and Europe [41-44].

The erythrocyte sedimentation rate has been correlated with blood haemoglobin in the diagnosis of tubal pregnancy by some workers[44, 41, 45], but others have failed to find any direct relationship[46, 47]. The erythrocyte sedimentation rate is known to be raised in cases of internal haemorrhage, and would, therefore, appear accelerated in the event of ruptured ectopic pregnancy. However, the evidence with accelerated or slow E.S.R. as a pointer to the differential diagnosis of tubular pregnancy is not conclusive.

Congenital malformation has been reported in infants born to malnourished women in some parts of the Tropics. While some of these cases are probably induced by mutant genes, animal experiments [52-54] suggest that dietary deficiencies of vitamins in the mothers, during gestation, may be the major cause. Non-hereditary phenocopies[50] have been produced in rats by withholding riboflavin from the pregnant mothers[48]. Pteroylglutamic acid deprivation has yielded similar results[49]. The appearance of the mutant genes themselves may be the outcome of deranged biochemical processes accompanying serious and prolonged malnutrition in pregnancy. The incidence of malformations of the central nervous system increased in malnourished German children during the 2nd World War, and a cause-and-effect relationship has been postulated between the two[51].

The eradication of endemic cretinism, in Europe, by the addition of iodine to table salt and by other dietary measures is a good indication of the relation between maternal nutrition and malformation of the off-spring[55]. Yet, this congenital disease is still widespread in some parts of Africa and Asia.

That the congenital malformations arising from vitamin deficiency in the pregnant women are due to the lack of specific chemical compounds is indicated by the fact that the morphological characteristics of vitamins A and D are easily distinguishable from those due to nutritional inadequacy of riboflavin, folic acid, and vitamin B_{12}[55].

Unfortunately, most of the biochemical studies in this field have been carried out on laboratory animals and there are not enough statistical data on human beings to make the animal experiments really meaningful.

Electrolyte Balance

One of the major effects of the physiological adaptations to pregnancy is on the body water content, and therefore in the electrolyte balance. LUND[59] found a plasma volume rise, varying from 14—121% in healthy pregnant women. VEREL, BURY & HOPE[56] found an increase in plasma volume in all of 13 pregnant women examined over the mean values for 8 non-pregnant English women. But ^{32}P-labelled red cell dilution showed an increase in the red cell volume in only 2 of the 13 women. The ratio of body haematocrit over venous haematocrit ranged much wider in the pregnant women (0.87—1.08) than in the non-pregnant subjects (0.9—0.95). It is therefore concluded that the measurement of blood volume changes in pregnancy by either of the two techniques, Evans Blue and ^{32}P dilution, is open to serious errors.

Transmineralization of sodium and potassium in normal pregnancy and in eclampsia and other complications have been studied in the uterus muscle, blood serum and erythrocytes by HEROLD[57]. The mean values obtained with 20 non-pregnant and 25 pregnant women were as follows: K 20.6 mg/100 ml serum and Na 318.5 mg/100 ml serum for the non-pregnant; K 20.3 mg/100 ml and Na 276.8 mg/100 ml for the pregnant. The red blood cell results were, for the non-pregnant, K 421.3 mg%, Na 116.7 mg%. For the pregnant the results were 391.9 mg% and 139.9 mg% respectively. Thus, while the serum K: Na ratio rose from 0.6652 in the non-pregnant to 0.726 in late pregnancy, the corresponding change in the red cells was from 3.604 to 2.807. The variation reported for K: Na in uterus muscle was 1.886: 1.157. In toxaemia the trend of K/Na was in the same direction as during normal pregnancy except that it was more marked. On account of the increase in the total water content of the body and the fact that the electrolyte figures were not given in milliequivalents, the indication of increased general metabolism in pregnancy is not precise.

The observations of DE ALVAREZ, SMITH & BRATVOLD[58] on the influence of dietary sodium intake on water, electrolyte, and nitrogen balance in pregnancy toxaemia has relevance to pregnancy disorders in tropical people who live on a low protein diet. And it implies that protein deficiency may worsen the electrolyte imbalance of pre-eclampsia. Obviously, more work on the relationship between body water loss in a condition of high ambient temperature and low-protein malnutrition is called for.

At term the mean values for total serum protein and for albumin for 36 pregnant women of the Kwango tribe of the Congo were found by HOLEMANS[60] to be below the average for non-pregnant women in the district. The total serum proteins and albumin steadily rose in value during the subsequent lactation, extending over a two year period, until it attained the normal average figures. As these women were all on a low-protein diet, the haemo-dilution imposed by pregnancy is easily explained, and it must have been accompanied by a deviation from the normal electrolyte balance.

It is difficult to understand why the albumin and total protein of the blood serum should progressively rise during lactation when at least 0.61 g of protein N was secreted in the milk daily, unless the dietary intake improved during breastfeeding.

Although the total serum proteins of a group of twelve pregnant Finnish women were lower than the normal values for their non pregnant counterpart, FURUHJELM[61] found that the total serum protein concentration of the umbilical blood of the infants was consistently lower than that of the maternal blood, averaging 5.8 and 6.9 g per 100 ml. In the course of the pregnancies the serum albumin of the women was reduced, but the beta globulin concentration increased. The resulting changes in the buffering capacity of the blood were associated with the observed increase in the blood sedimentation rate of 32 pregnant women, but there was not a statistical correlation between the two.

A recent estimation of water transfer from the amniotic fluid to the foetus, in which deuterium oxide was injected into the amniotic sac during caesarian sectioning, suggested that the rate of exchange between mother and foetus is about 2—3 l/hour. This represents a turnover of amniotic fluid of about half a litre an hour, of which at least 25% is via the foetus[62]. This study gives a good indication of the intimate relationship between the body fluid proteins of the mother and foetus and it stresses the dynamic principle underlying the alterations of serum protein concentration in pregnancy. Similar body water changes have been reported by other workers[63, 64].

Very little work appears to have been done in the Tropics on the production of hydrochloric acid in the gastric juice during pregnancy. KETHAIN & BHENDE[65] in India studied a small number of cases and found reduced free acidity but achlorhydria was only observed in one case. There was no correlation between gastric acidity and blood haemoglobin concentration. In this respect their results agree with those of WATSON[66] and BETHELL[67] but not fully with the findings of STRAUSS & CASTLE[68], according to whom achlorhydria occurs in about three-quarters of pregnant American women studied.

The connection, if any, between the carbon dioxide combining power of blood plasma and the low level of free acid in the gastric juice of pregnant women in the tropics will probably repay investigation. BOUTOURLINE-YOUNG & BOUTOURLINE-YOUNG[69] have determined the alveolar carbon dioxide levels in pregnant, parturient and lactating mothers in America and their observations could be used as a baseline in any study of the part played by HCO_3 ions in the maintenance of electrolyte balance in pregnant women of Asia and Africa.

Mineral Metabolism

A great deal of interesting work has been done on the metabolism of calcium during pregnancy, particularly in domestic animals. Much of this has been reviewed by HILL[70] and will not be repeated here. More recently FEASTIN, HANSARD, OUTLER & DAVIS[71] maintained rats on a diet containing 1.38% Ca and 0.96% P to the age of 12—15 months when they were mated. They were then given by mouth single doses of ^{45}Ca at various stages of pregnancy and later sacrificed for the determination of radioactive calcium in the foetus and mothers. The result of this experiment indicated that a considerable amount of calcium was absorbed into the maternal blood stream and transferred across the placenta to the foetus, in half an hour. However, pre-formed calcium, from stores in the mother's body, was transferred as well. From the 14th to the 22nd day of pregnancy, the amount of ^{45}Ca transferred across the placental barrier to the foetus bore a direct relationship with the stage of pregnancy.

In terms of the malnutrition of pregnant women in the Tropics, this observation implies that, provided an adequate intake of calcium and phosphorus is maintained during pregnancy, a disastrous drain of the natural stores of calcium can probably be averted. In many parts of Africa peasant women forage on chalky earth, particularly in the later stages of pregnancy. In view of the low concentration of dietary calcium, to which reference has been made in Chapter III, it is almost certain that there is usually a great demand for extra dietary calcium during pregnancy in Africans. The chalk ingested would therefore be partitioned between repleting the mother's stores and directly supplying the foetus in the manner described by FEASTER and his colleagues[71].

Even in as short an interval as three days a dietary supplement containing 375 mg Ca, 188 mg P, 600 I.U. vitamin D, iron and vitamins A, C, and B complex has been shown to increase the retention of calcium in four women, in the third trimester of pregnancy[72].

By estimating the levels of protein, calcium and phosphorus in the blood serum of 50 pregnant women for five weeks in the third trimester, HARDY[73] was able to demonstrate the precise effect of different kinds of dietary supplementation of calcium. The best result was obtained with a vitamin and mineral supplement containing calcium lactate. Both the total and ionic Ca, as well as protein, rose during the experimental period. The phosphorus level remained unchanged. As compared with the 10 untreated control group receiving no dietary supplement, the batch of 10 women who were given the usual vitamin and mineral preparation, containing dicalcium phosphate and vitamin D, fared worse. The percentage decreases in their total and ionic serum calcium was greater than in the controls. But the phosphorus and protein contents rose.

In 1941 SCHULTZE[74] showed that the cytochrome oxidase activity of bone marrow depended on the intake of copper in the diet. It had previously been established that the synthesis of cytochrome A required normal copper metabolism in animal tissue[75]. And it seems clear that the need of copper in the synthesis of the haem nucleus is the specific effect of this element on iron metabolism which results, clinically, in a microcytic hypochromic anaemia in many species of mammals[76].

GALLAGHER, JUDAH & REES[77] have made comprehensive studies on the relation of copper to the level of enzyme activities in copper-deficient rats. Weanling rats were given a basal milk diet made deficient by treatment with hydrogen sulphide. Vitamins, iron and other minerals were added to the diet of the experimental animals, while the controls got daily 50 micrograms Cu in addition. The copper deficient animals showed reduced growth as compared with the controls. Anaemia and death followed in due course. The tissues of both control and test animals were assayed for 14 different enzymes, including those of the Krebs tricarboxylic acid cycle, cytochrome oxidase, catalase, transmethylase and choline oxidase. It was found that only the terminal enzyme of biological oxidation (cytochrome oxidase) showed a marked reduction in the copper-deficient animals.

There is strong evidence that copper may act as a growth stimulant in mammals, but opinion in this matter is not unanimous. Following the observation of BARKER, BRAUDE, MITCHELL & CASSIDY[78] that a mineral supplementation of standard pig rations, which added small amounts of copper sulphate, increased the growth rate of the animals, other workers in different parts of Britain[79] have obtained similar results by dietary supplementation with 250 parts per million copper. However, CARPENTER[80] found that the addition of copper sulphate to a practical ration had no effect in stimulating growth in lambs, while with one-tenth the concentration of additional copper used by BOWLER and his colleagues[79], he obtained approximately the same increase in growth rate as did the British workers. It is difficult to say whether the growth-stimulating effect of dietary copper in pigs is due to its anti-microbial action or its role in stimulating enzyme action.

Great caution is necessary in interpreting the results of these animal experiments in terms of human nutrition.

It is obviously of interest to consider the findings of MIRZAKARI-MOV[19] who estimated the concentration of copper in the blood of a group of 62 women at various stages by gestation. The results which are given in Table XVIII indicate a progressive increase in the blood content of the element during the course of pregnancy.

Table XVIII

Mean Values and Ranges of Blood Copper During Pregnancy

Number of Subjects	Month of Pregnancy	Range (mg Cu/100 ml)	Mean
4	2	86.6 – 119.2	105.5
5	3	105.6 – 134.6	119.6
5	4	110.8 – 145.0	128.2
6	5	104.6 – 151.6	128.9
9	6	114.8 – 169.2	137.2
9	7	133 – 181.4	152.2
11	8	143 – 191.4	168.8
8	9	151.1 – 215.3	185.2
5	10	153.2 – 211.6	178.5

Extracted from the results of M. G. MIRZAKARIMOV[19].

That this increase in the blood level of copper is obtained at the expense of the tissue stores is suggested by the observation of the same author that the concentration of the mineral in the tissues of dead pregnant women was lower than in the non-pregnant. An indirect support of this conclusion is the fact that the intake of 5 mg copper daily, as sulphate, had no effect on the level of blood copper during the second half of pregnancy.

Since the concentration of Cu in the blood returned to the non-pregnant normal level two months after parturition, it seems very likely that the level of the blood Cu during pregnancy is a measure of the increasing needs of the growing foetus, but that this need cannot be met directly by dietary intake. It thus appears that copper metabolism is bound up in some intricate manner with foetal growth.

There is no direct evidence that copper deficiency has a limiting role in the development of the foetus in tropical women, although the severe anaemia of pregnancy from which some of these women suffer suggests that it might. It would be interesting to know whether the addition of catalytic amounts of copper to the diet of anaemic African women reduces their intractibility to treatment with iron. If parallel enzymatic studies on the oxidase systems are undertaken, real progress towards understanding the role of copper in pregnancy may be made.

Since a high incidence of abortion has been reported in many parts of the Tropics where malnutrition prevails[81] and because of the interrelationship of the activity of the thyroid gland and that of the gonads, it is logical to expect augmentation of the former to lead to more efficient function of the latter in preventing spontaneous abortion. However, SINGH & MORTON[82] found that thyroid medication had no effect on four women, three of whom had a history of previous abortions. Values of protein-bound iodine in the blood of 25 normal pregnant women ranged between 6.99 and 7.58 µg per 100 ml, and after delivery fell to 6.58 µg in six weeks. The range was 5.5—8.9 µg during pregnancy in the 26 women who aborted. Only in one women, who aborted at 7 months, a high protein-bound iodine value (11.2 µg per 100 ml) was found.

By giving 10—15 µC[131]I-thyroxine to 15 pregnant women at varying periods before delivery and estimating the organically bound serum radio-active iodine in both mother and foetus it was found by GRUMBACH & WERNER[83] that the maternal: foetal ratio fell from about 40, after two hours, to the order of 8 after 18 hours. By the 169th hour after the injection, the ratio had decreased to approximately 2. Since sodium iodide was given to the mother immediately before the administration of ^{131}I-thyroxine and maintained throughout the experiment, it was considered that both the maternal and foetal thyroids were saturated with the drug during this period, thus preventing the accumulation there of labelled ^{131}I or thyroxine. It may be deduced from this experiment that placental transfer of the thyroid hormone is slow, although a considerable amount of transportation takes place.

This slow rate of trans-placental transfer of the thyroid hormone is the probable explanation for the unexpected result of the experiments of (a) SINGH & MORTON[82], to which reference has already been made, and (b) LYBECK[84] who failed to obtain evidence of transmission of ^{131}I labelled tri-iodothyronine from mother to foetus 2 to 4 hours after the injection of 10—15 µC of the isotope into pregnant female guinea pigs. The experiment of GRUMBACH & WERNER with humans extended over a much longer period, and they found an appreciable transfer of ^{131}I tri-iodothyronine across the placenta, as reflected in the fall in the maternal/foetus ratio of organically-bound serum radioactive iodine.

Therefore, it may well be that if the level of thyroxine-like substances in the blood of the mother is to confer an advantage for the survival of the foetus, fortification with less complex iodine compounds (to which the placenta is more permeable) must be undertaken.

The place of the dietary iron in tropical malnutrition has been described earlier in connection with anaemia of pregnancy. A few words remain now to be said on the transport of the element. The distribution of the extra 650 mg iron required during the second half of pregnancy is made between the foetus (250 mg) [75, 86], the placenta (50—100 mg)[87] and the mother's additional haemoglobin (300 mg)[88, 89]. The requirement of the foetus and placenta does not need to be met from dietary sources only since the maternal stores may amount to about 1500 mg at the beginning of pregnancy[90], but the fact that the proportion of iron absorbed from a given dose increases markedly (e.g. 3.7 mg from 9 mg of labelled iron) in the last three months[91] suggests that the problem of trans-placental transport of the metal is a complex one[98, 99].

In an investigation of embryonal iron turnover, EHRENSTEIN & HEVESY[106] injected ^{59}Fe into pregnant rabbits exposed to X-rays and determined the proportion of the administered dose which became localized in the livers of both mother and foetus. Although the weight range of the livers of the foetuses was 1.5g—5.9g while that of the mothers was 86—170 g, up to 33% of the ^{59}Fe given was found in foetal liver. This was much more than in the mothers' livers. Exposure to X-rays also increased the ratio of specific activity of foetal to maternal circulating haemoglobin. Haemopoiesis declined in both the mothers and their 20 days old embryos. Of the amount of radioactive iron transported across the placenta to the foetus after X-ray treatment a day before the ^{59}Fe injection, a smaller fraction was incorporated into the nuclear fraction of the liver than in the controls in which the specific activity of the haemoglobin in the nuclear fraction of the foetal liver was twice that of the circulating Hb. The dynamic aspect of the transport problem is stressed by the fact that the specific activity of Hb, in the nuclear fraction of the maternal liver, was only a fourth of that of circulating Hb. The sequence appears to be less metabolic activity of Hb iron in the liver of the mother, accompanied by a high turnover rate in the blood and a much higher one in the glandular tissues of the foetus.

VENTURA & KLOPPER[92, 93] have put forward an explanation of the "vastly increased turnover of iron in late pregnancy" in terms of a raised iron binding capacity of the blood. But since the iron-binding property of serum is associated with the beta globulin transferrin, the observation of GERRITSEN & WALKER[94], who found increased iron-binding capacity in Bantu women who showed no decrease in serum iron or haemoglobin, must recommend raised plasma globulin[95], rather than the saturation of iron-binding capacity[96], as a more reasonable interpretation of the experimental data.

Kidney Function

It is well known that the function of the kidney changes in human beings in conditions of gross malnourishment (NAKAZAWA & KUSAKARI[100], MOLLISON[101]. In starvation oedema there is a striking reduction of kidney function. When pregnancy complications are imposed on a general condition of malnourishment, the effect on kidney function may be very marked.

In the Tropics, the effect of dietary intake of salt on kidney function is of the utmost importance in view of the high sweat-rate and high ambient temperatures.

Animal experiments suggest that food intake affects the body's response to the adrenal hormones and it can be shown that in adrenal insufficiency in rats partial starvation has a normalising effect by reducing the retention of Na+, K+, and Cl− especially in the genesis of nutritional oedema. ROBINSON & FARR[102] showed many years ago that a correlation is possible between the amount of the anti-diuretic hormone in the urine and the development of the oedema. RALLI, ROBSON, CLARKE & HOAGLAND[103] are of the opinion that the development of ascites in the liver disease is due to the inability of the liver to inactivate pituitary anti-diuretic hormone. GILLMAN & GILLMAN[81] include hypothyroidism as one of the factors leading to the retention of sodium chloride in nutritional oedema.

It is almost certain that the reduction in the excretion of salt in pregnancy, which is known to occur, is part of a complex inter-relationship of hormonal secretions. Within limits we can conceive it as an adaptation designed to ensure the wholesomeness of the aqueous environment in which a healthy foetus grows.

It also seems fairly clear that aldosterone metabolism is not impaired in pregnancy and that the reported elevated levels of cortisol may be due to increase in the concentration of cortisol-binding protein in plasma[105].

Much ingenuity will be required to separate the metabolic strands by which electrolyte balance and kidney function are bound in that whirlpool of hormonal activity—pregnancy.

REFERENCES

1. MACY, I. G. & MACK, H. C., 1954. *Amer. J. Obstet. Gynec.* **68**, *131.*
2. HUMMEL, F. C., STERNBERG, H. R., HUNSCHER, H. A. & MACY, I. G., 1936. *J. Nutrit.* **11**, *235.*
3. HUNSCHER, H. A., G. C., ERICKSON, B. N. & MACY, I. G., 1935. *J. Nutrit.* **10**, *579.*
4. HOGAN, A. G., 1953. *Ann. biochem. Rev.* **22**, *299.*
5. BURKE, B. S. & STUART, H. C., 1948. *J. Amer. med. Ass.* **137**, *119.*
6. TOVERUD, K. U., STEARNS, G. & MACY, I. G., 1950. *Nat. Res. Counc. Bull.* **123**.
7. McGANITY, W., CANNON, R., BRIDGFORTH, E., MARTIN, P., DENSEN, P. M., NEWBILL, A., McCALLANS, S., CHRISTIE, A., PETERSON, J., J. & DARBY, W. J., 1954. *Amer. J. Obstet. Gynec.* **67**, *501.*
8. DARBY, W. J., CANNON, R. O. & KASER, M., 1948. *Obstet. Gynec. Surv.* **3**, *704.*
9. MACK, H. C., KELLY, H. C. & MACY, I. G., 1956. *Amer. J. Obstet. Gynec.* **71**, *577.*
10. MOYER, E. Z., MACY, I. G., MACK, H. C., DI LORETO, P. C. & PRATT, J. P., 1954. Nutritional Status of Mothers and Their Infants. Childrens Fund of Michigan, Detroit.

114

11. PLATT, B. S., 1954. *Lancet* **2**, *929.*
12. WOODRUFF, A. W., 1960. Proc. 5th Congr. trop. Med. and Malaria.
13. WATKINS, S., 1955. Personal Communication.
14. EDOZIEN, J., ADENIYI-JONES, C. & WATSON-WILLIAMS, E. J., 1961. *Brit. med. J.* **i**, *333.*
15. MACY, I. G. & MACK, H. C., 1952. Physiological Human Reproduction. Children's Fund of Michigan, Detroit.
16. MACK, H. C., 1955. The Plasma Proteins in Pregnancy. Charles C. Thomas, Springfield, Ill.
17. SCRIMSHAW, N. B., GUZMAN, M. & DE LA VEGA, J. M., 1951. *Amer. J. trop. Med.* **31**, *163.*
18. PARVIAINEN, S., SOIVA, K. & EHONROOTH, C. A., 1949. *Acta obstet. gynec. scand.* **29**, *186.*
19. MIRZAKARIMOV, M. G., 1958. *Akusherstvo i ginekologii.* (Moskva). **34**, 4, *90.*
20. WILLS, L. W., 1931. *Brit. med. J.* **1**, *1059.*
21. WILLS, L. W., 1953. *Indian J. med. Res.* **21**, *669.*
22. MANSON-BAHR, P. E. C., 1951. *Trans. R. Soc. trop. Med. Hyg.* **44**, *555.*
23. NAPIER, L. E., 1940. *Indian J. med. Res.* **27**, *1009.*
24. WOODRUFF, A. W., 1951. *Brit. med. J.* **2**, *1415.*
25. WOODRUFF, A. W., 1954. In: Malnutrition in African Mothers, Infants and Young Children. H. M. S. O., London.
26. WOODRUFF, A. W., 1955. *Brit. med. J.* **1**, *1297.*
27. BURKE, B. S., HARDING, V. V. & STUART, H. C., 1943. *J. Pediat.* **23**. *506.*
28. DIECKMANN, W. J. & WEGNER, C. R., 1934. *Arch. intern. Med.* **53**, *71.*
29. DIECKMANN, W. J. & WEGNER, C. R., 1934. *Arch. intern. Med.* **53**, *188.*
30. STURGIS, C. C., 1948. Hematology, Charles C. Thomas, Springfield, Ill.
31. DIECKMANN, W. J., 1952. The Toxemia of Pregnancy. The C. V. Mosby Co. St. Louis.
32. LAWSON, J., 1957. Personal Communication.
33. HENDERSON, D. N. & BEAN, J. L., 1950. *Amer. J. Obstet. Gynec.* **59**, *1225.*
34. WARE, H. H. & WINN, W. C., 1946. *Southern med. J.* **39**, *44.*
35. DAVIS, J. D. & MALLOY, R. R., 1948. *J. Nat. med. Assoc.* **40**, *3.*
36. FALL, F. H., 1950. *Surg. Clin. North Amer.* **30**, *207.*
37. MACFARLANE, K. T. & SPARLING, D. W., 1946. *Amer. J. Obstet. Gynec.* **51**, *343.*
38. WORD, B., 1951. *Surg. Gynec. & Obstet.* **92**, *333.*
39. JOHNSON, W. O., 1952. *Amer. J. Obstet. Gynec.* **64**, *1103.*
40. LAWSON, J., 1956. Clinical report of the Dept. Obst. & Gynec. of Ibadan Univ. Coll.
41. HU, I. F., 1950. *Carle Hosp. Chi. Bull.* **3**, *18.*
42. CAMPBELL, R. M., 1952. *Amer. J. Obstet. Gynec.* **63**, *54.*
43. BEACHAM, W. D., 1951. *J. Missouri med. Ass.* **48**, *629.*
44. CRAWFORD, E. & HUTCHINSON, H., 1954. *Amer. J. Obstet. Gynec.* **67**, *568.*
45. DOUGLAS, G. F., 1951. *J. int. Coll. Surgeons* **15**, *28.*
46. TE LINDE, R. W., 1946. Operative Gynecology. J. B. Lippincott Co., Philadelphia, Pa.
47. NOVAK, E., 1947. Gynecological and Obstetrical Pathology. W. B. Saunders, Co. Philadelphia.
48. WARKANY, J., NELSON, R. C. & SCHRAFFENBERGER, E., 1943. *Amer. J. Dis. Child.* **65**, *882.*
49. NELSON, M. M., ASLING, C. W. & EVANS, H. W., 1950. *Amer. Rec.* **106**, *309.*
50. GOLDSCHMIDT, R. B., 1938. Physiological Genetics. McGraw-Hill Co., New York.
51. EDITORIAL of *Lancet,* 1952 **263**, *977.*
52. NUTRITION REVIEWS, 1944 **2**, *297.*
53. NUTRITION REVIEWS, 1947 **5**, *89.*
54. NUTRITION REVIEWS, 1954 **12**, *189* and *348.*
55. WARKANY, J., 1955. *Nutrit. Rev.* **13**, *290.*

56. VEREL, D., BURY, J. D. & HOPE, A., 1956. *Clin. Sci.* **15** *1.*
57. HEROLD, L., 1955. *Arch. Gynäkol. Abt. Städt. Krankenhaus Düsseldorf,* **187**, *388.*
58. DE ALVAREZ, R. R. & SMITH, E. K., 1956. *Amer. J. Obstet. Gynec.* **72**, *562.*
59. LUND, C. J., 1951. *Amer. J. Obstet. Gynec.* **62**, *947.*
60. HOLEMANS, K., 1955. *Ann. Soc. belge Méd. trop.* **35**, *29.*
61. FURUHJELM, U., 1956. *Ann. Paediat. Fenn.* **2** (5), *75.*
62. GRAY, M. J., NESLEN, E. D. & PLENTL, A. A., 1956. *Proc. Soc. exp. Biol. Med.* **92**, *463.*
63. SEITCHIK, J. & ALPER, C., 1956. *Amer. J. Obstet. Gynec.* **71**, *1165.*
64. HALEY, H. B. & WOODBURY, F. W., 1956. *Surg. Gynec. & Obstet.* **103**, *227.*
65. KETHAIN & BHENDE, Y. M., 1952. *Indian J. med. Res.* **40**, *387.*
66. WATSON, H. G., 1938. *Amer. J. Obstet. Gynec.* **35**, *106.*
67. BETHELL, F. H., 1948. *J. Lab. clin. Med.* **33**, *1477.*
68. STRAUSS, M. B. & CASTLE, W. B., 1932. *Amer. J. med. Sci.* **184**, *655.*
69. BOUTOURLINE-YOUNG, H. & BOUTOURLINE-YOUNG, E., 1956. *J. Obstet. Gynaec. Brit. Empire,* **63**, *509.*
70. HILL, R., 1954. *Agric. Progr.* **29**, *98.*
71. FEASTER, J. P., HANSARD, S. L., OUTER, J. C. & DAVIS, G. K., 1956. *J. Nutrit.* **58**, *399.*
72. NEWMAN R. L., 1956. *Obstet. Gynec.* **8**, *561.*
73. HARDY, J. A., 1956. *Obstet. Gynec.* **8**, *565.*
74. SCHULTZE, M. O., 1941. *J. biol. Chem.* **1**, **138**, *219.*
75. COHEN, S. E. & ELVEHJEM, C. A., 1934. *J. biol. Chem.* **107**, *97.*
76. NUTRITION REVIEWS, 1953, **11**, *336.*
77. GALLAGHER, D. H., JUDAH, J. D. & REES, K. R., 1956. *Proc. Roy. Soc.* **B. 145**, *134.*
78. BARBER, R. S., BRAUDE, R., MITCHELL, K. G. & CASSIDY, J., 1955. *Chem. Ind.* *601.*
79. BOWLER, R. J., BRAUDE R., CAMPBELL, R. C., CRADDOCK-TURNBULL, J., FIELDSEND, H., GRIFFITHS, E., LUCAS, I. A., MITCHELL, K. G., McKALLS, N. & TAYLOR, J., 1955. *Brit. J. Nutrit.* **9**, *358.*
80. CARPENTER, L. E., 1957. *Annual Rep. Harwell Inst.* 21.
81. GILLMAN, T. & GILLMAN, T., 1951. Perspectives in Human Malnutrition. Grune and Stratton, New York. p. *421.*
82. SINGH, B. P. & MORTON, D. G., 1956. *Amer. J. Obstet. Gynec.* **72**, *607.*
83. GRUMBACH, M. M. & WERNER, S. C., 1956. *J. clin. Endocrinol.* **16**, *1392.*
84. LYBECK, H., 1956. *Acta physiol. scand.* **37**, *215.*
85. IOB, V. & SWANSON, W. W., 1938. *J. biol. Chem.* **124**, *263.*
86. WIDDOWSON, E. M. & SPRAY, C. M., 1951. *Arch. Dis. Childh.* **25**, *205.*
87. RECHENBERGER, J., 1955. *Dtsch. Z. Verdauungs u. Stoffwechselkr.* **15**, *12.*
88. MOLLIN, D. L. & ROSS, G. 1952. *J. clin. Path.* **5**, *129.*
89. BERLIN, N. I., GOETSCH, C., HYDE, G. M. & PARSONS, R. J., 1953. *Surg. Gynec. Obstet.* **97**, *173.*
90. HACKINS, D., STEVENS, A. R., FINCH, S. & FINCH, C. A., 1952. *J. clin. Invest.* **31**, *543.*
91. HAHN, P. F., CAROTHERS, E. L., DARBY, W. J., MARTIN, M., SHEPPARD, C. W., CANNON, R. D., BEAM, A. S., DENSEN, P. M., PETERSON, J. C. & McCLELLAN, G. S., 1951. *Amer. J. Obstet. Gynec.* **61**, *477.*
92. VENTURA, S. & KLOPPER, A., 1951. *J. Obstet. Gynaec. Brit. Empire* **58**, *173.*
93. VENTURA, S. & KLOPPER, A., 1951. *S. Afr. med. J.* **25**, *969.*
94. GERRITSEN, T. & WALKER, A. R., 1954. *J. clin. Invest.* **33**, *23.*
95. HYTHE, F. E. & DUNCAN, D. L., 1956. *Nutrit. Abstr. and Rev.* **26**, *855.*
96. LAURELL, C. B., 1947. *Acta physiol. scand.* **14**, Suppl. 46.
97. PRIBILLA, W. & GEHMANN, G., 1956. *Folia Haemat. Frankfurt,* **1**, *23.*
98. NYLANDER, G., 1954. *Acta Soc. med. Upsalien.* **59**, *372.*
99. SCOTT, J. M., 1954. *J. Obstet. Gynaec. Brit.-Empire* **61**, *641.*
100. NAKAZAWA, F. & KUSAKARI, H., 1930. *Tohoko J. exp. Med.* **16**, *321.*

116

101. Mollison, P. L., 1946. *Brit. med. J.* **1**, *4*.
102. Robinson, F. H. & Farr, L. E., 1940. *Ann. intern. Med.* **14**, *42*.
103. Ralli, E. P., Robson, J. S., Clarke, D. & Hoagland, C. L., 1945. *J. clin. Invest.* **24**, *316*.
104. Woodruff, A. W., 1961. In: Recent Advances in Human Nutrition. Churchill, London.
105. Nutrition Review, 1959. 17, *208*.
106. Ehrenstein, G. V. & Hevesy, G., 1956. *Acta physiol. scand.* **38**, *184*.

NAMES INDEX

118

120

Stearns, G., 22, 25, 45
Steege, H., 63
Stetten, de, W., 27
Stoa, K.F., 63
Stransky, E., 45, 46
Strauss, M.B., 107
Stringer, D., 45
Stuart, H.C., 104
Subrahmanyan, V., 72
Sung, C., 24
Swaminathan, M., 72
Sydow, G., 23

Taliaferro, W.H., 86
Tanguy, F., 40
Tappel, A.L., 63
Teitelbaum, H., 29
Terzian, L., 87
Thamhipillai, 94
Thompson, H., 89
Thompson, M.D., 56, 57, 58, 67, 90
Todd, W.R., 23
Tomarelli, R.M , 58
Toverud, G., 36
Toverud, K.U., 22
Tremolieres, J., 56
Trowell, H.C., 28, 40, 56, 62, 63, 64, 66, 90
Turnbull, E.P., 15
Turnbull, J.M., 65
Turner, R., 66

Uga, Y., 42

Van Der Sar, A., 56
Van Sande, M., 85
Veghelyi, P.V., 90
Venkatachalam, P.S., 66, 69

Ventura, S., 112
Verel, D., 106
Vicente, C., 46
Vint, F.W., 30

Walker, A.R.P., 70, 112
Walker, J., 10
Walker, N.F., 20
Walker, R.A., 36
Waller, H., 36
Walters, J.H., 42, 91
Waterlow, J., 26, 42, 56, 89, 91
Watson, H.G., 107
Watson-Williams, E.J., 25
Watt, A., 10
Weichselbarem, T.E., 91
Weinstock, M., 43
Welbourn, H.F., 18, 45
Werner, G., 89
Werner, S.C., 111
Wetzel, N.C., 67
Widdowson, E.M., 89
Williams, L.L., 45
Williams, R.D., 74
Wilner, J.S., 94
Wintrobe, M.M., 83
With, T.K., 44
Wohl, M.G., 66
Wolffson, D., 90
Woo, T.T., 24
Woodruff, A.W., 25, 37, 101, 102, 103
Wooley, D.W., 68
Wright, E.J., 22

Yudkin, S., 27

Zaimis, J.E., 30
Ziemann, H., 86

SUBJECT INDEX

Acid Phosphatase, 89
Agidi-baba, 37
Alkaline Phosphatase, 65, 89
Alkalosis, 54
Amniotic Fluid transfer, 107
Anaemia, 64
Anti-malarials, 86
Antidiuretic hormone, 113
Ascorbic acid oxidase, 69

"Barts" haemoglobin, 20
Basal metabolism, 73
Blood copper in pregnancy, 110
Blood electrolytes, 55
Body water, 53
Breast milk composition, 38, 39
Bromsulphthalein excretion, 27

45 Ca, 93, 95, 108
Ca/P ratio, 93
Calcium requirement, 43
Caloric requirement, 47
Calorie intake, 71
Carbonic anhydrase, 19
Cassava, 76
Choline esterase, 89
Colostrum, 35
Complementary feeds, 47
Copper deficiency, 109
Cortisone, 91
Cretinism, 31
Cytochrome oxidase, 109

Decalcification, 94

Electrocardiograph, 65
Electrolyte imbalance, 52, 107
Endogenous calcium, 95
Enzyme mononucleotide amino acid, 83
Enzymes, 26, 78, 109
Erythrocyte Sedimentation Rate, 105

Foetal Haemoglobin, 18
Folic acid, 68
Food and Agricultural Production, 13

Galactose tolerance, 73
Gastric function, 55

Haematological picture in pregnancy, 103
Haemodilution, 104
Haemoglobin, 86, 88, 104, 112
Haptoglobulin negatives, 20
Hepatic fibrosis, 91
Hollow hunger, 99
Hyperbilirubinemia, 27
Hypothyroidism, 31

Immunity to Malaria, 86
Infant feeding Schedule, 48, 49
Infant Mortality, 9
Iodine tracers, 111
Iron, 111, 112

Keratin, 77
17-Ketosteroid, 29, 69, 90
Kwashiorkor, 10, 55, 57, 62, 90

Liver dysfunction, 26, 30

Macrocytic anaemia, 74
Malaria, 15
Megaloblastic anaemia, 102
Methyl Syntheses, 68
Milk amino acids, 41
Mortality rates, 16

Natural Vegetation, 7
Nutrient intakes of Pregnant women, 100

Oedema, 113
Oestrogen, 29

32P red-cell dilution, 106
Pancreas, 92
Pancreatic function, 56
Pennisetum, 84
Phenocopies, 105
Physiological anaemia, 25
Phytic acid, 93
Pituitary gland, 30
Placenta, 24
Plantain, 73
Plasma proteins, 84
Plasmodium, 85
Prematurity, 100

(283)